Studies in Big Data

Volume 125

Series Editor

Janusz Kacprzyk, Polish Academy of Sciences, Warsaw, Poland

The series "Studies in Big Data" (SBD) publishes new developments and advances in the various areas of Big Data- quickly and with a high quality. The intent is to cover the theory, research, development, and applications of Big Data, as embedded in the fields of engineering, computer science, physics, economics and life sciences. The books of the series refer to the analysis and understanding of large, complex, and/or distributed data sets generated from recent digital sources coming from sensors or other physical instruments as well as simulations, crowd sourcing, social networks or other internet transactions, such as emails or video click streams and other. The series contains monographs, lecture notes and edited volumes in Big Data spanning the areas of computational intelligence including neural networks, evolutionary computation, soft computing, fuzzy systems, as well as artificial intelligence, data mining, modern statistics and Operations research, as well as self-organizing systems. Of particular value to both the contributors and the readership are the short publication timeframe and the world-wide distribution, which enable both wide and rapid dissemination of research output.

The books of this series are reviewed in a single blind peer review process.

Indexed by SCOPUS, EI Compendex, SCIMAGO and zbMATH.

All books published in the series are submitted for consideration in Web of Science.

Douglas G. Woolford · Donna Kotsopoulos ·
Boba Samuels

Editors

Applied Data Science

Data Translators Across the Disciplines

Editors
Douglas G. Woolford
Department of Statistical and Actuarial
Sciences
University of Western Ontario
London, ON, Canada

Donna Kotsopoulos Ⓘ
Faculty of Education
University of Western Ontario
London, ON, Canada

Boba Samuels
Faculty of Kinesiology and Physical
Education
University of Toronto
Toronto, ON, Canada

ISSN 2197-6503   ISSN 2197-6511   (electronic)
Studies in Big Data
ISBN 978-3-031-29939-1   ISBN 978-3-031-29937-7   (eBook)
https://doi.org/10.1007/978-3-031-29937-7

This Springer imprint is published by the registered company Springer Nature Switzerland AG
The registered company address is: Gewerbestrasse 11, 6330 Cham, Switzerland

Preface

People with data literacy proficiency—those identified as "data translators"—are in high demand in the workplace. In organizations of all sizes and scopes, there has been an explosion in the need and interest to use data to guide action. This evidence-based decision-making results in growing engagement in data science and analytics. Large and complex data sets are more readily available and even the public is using data to guide everyday decisions. Thus, data acumen is required across a broad range of fields and skill levels, but how do we foster its development?

Historically, data science instruction has focused on statistical, computing, and other technical competencies. In contrast, many users of data and data modeling methods may not be considered "data scientists", but are required nevertheless to work with data to address disciplinary problems and communicate data-driven solutions effectively for specific audiences. An enduring challenge is educating and training individuals who may not have a traditional data science background to become data translators.

This edited volume is a collection of discipline-specific examples of data-driven solutions undertaken by both data scientists and non-data scientists across a wide variety of fields. Contributions emphasize effective application of data modeling methods and communication while simultaneously highlighting the process of producing an effective disciplinary solution. The chapters illustrate a broad approach to developing data translators across a variety of fields, such as education, health sciences, natural sciences, politics, economics, business and management studies, social sciences, and humanities. Our challenge presented to contributing authors was to illustrate effective data translation in practice, while sharing pedagogical approaches that might be important for supporting the development of the non-traditional data translator.

The chapters are a collection of case studies that illustrate how data is used and translated in various disciplinary contexts. Authors share their data science solution lifecycle to illustrate data use within a discipline, approaches to translating and communicating results, and pedagogical approaches to developing effective data translators.

In Chapter "Translating Science into Actionable Policy Information—A Perspective on the Intergovernmental Panel on Climate Change Process", Zwiers and Zhang describe the ways in which the Intergovernmental Panel on Climate Change's (IPCC) six climate change assessments have evolved over its 35-year history. The authors show how the level of confidence in the understanding of the causes of climate change has evolved over time using careful and consistent calibrated language, resulting in the development of options for mitigation. The chapter effectively makes the case for shared language as an important feature of data translation.

Chapter "Data in Observational Astronomy" authors Barmby and Wong describe why astronomers collect data; how data are collected, processed, used and shared, both between astronomers and between astronomers and the public; unique aspects of astronomical data; and future challenges for telling the story of the universe with astronomical data. The chapter provides an insightful example of data translation across users of differing goals and expertise.

Chapter "Beyond Translation: An Overview of Best Practices for Evidence-Informed Decision Making for Public Health Practice" focuses on public health data translation. Schanzer and the NSERC/Sanofi/York Industrial Research Chair Disease Modelling Group (Arino, Asgary, Bragazzi, Heffernan, Seet, Thommes, Wu, and Xiao) describe how data can be used to develop public health guidelines, build consensus among healthcare officials/professionals, and by public end-users. These guidelines are typically written by multi-disciplinary committees with attention to the users' perspective. Data translation, thus, is distributed right to the end-user, the public.

Continuing with health data, Rahman in Chapter "Concern for Self-Health During the COVID-19 Pandemic in Canada: How to Tell an Intersectional Story Using Quantitative Data?" draws on intersectionality theory to illustrate how data can be used to make visible the inequalities of Canadians' concerns for self-health during the COVID-19 pandemic. Rahman's analysis of Statistics Canada COVID-19 Impacts Survey data 2020 demonstrates the importance of considering the interconnected nature of different characteristics of individuals when analyzing data in order to illuminate their experiences.

Chapter "Community-Based Participatory Research and Respondent-Driven Sampling: A Statistician's, Community Partner's and Students' Perspectives on a Successful Partnership" by authors Rotondi, Jubinville, McConkey, Wong, Avery, Bourgeois, and Smylie describes how community-based participatory research fully integrates statistical scientists into the research team with community partners. The outcome is a dynamic research relationship where both researchers and community partners are educated through and with shared data translation through their knowledge exchange. Like Chapter "Concern for Self-Health During the COVID-19 Pandemic in Canada: How to Tell an Intersectional Story Using Quantitative Data?", the authors provide a case for the optimal studying of hard-to-reach populations where ownership of study processes and results ensures research questions better reflect the community's priorities and needs.

Online and virtual learning has grown. The use of learning technologies and communication platforms has enabled unprecedented potential for the collection of detailed data on learners' actions in a virtual environment. Chapter "Operationalizing Learning Processes Through Learning Analytics" shifts to the discipline of education, where Patzak and Vytasek provide an overview of learning analytics that can be leveraged to enhance understanding and provide feedback about learning processes. The authors use research on procrastination to exemplify how different types of data can be used to interpret and operationalize learning processes.

In the context of management and business, the data analytics workflow starts with the definition of a business problem, and it ends with the creation of relevant interactive reports and dashboards that are then analyzed. The main instructional challenges when teaching students in this field are related to the cost of different software, students' low data literacy, and also their lack of technical skills. In Chapter "Improving Data Literacy in Management Education Through Experiential Learning: A Demonstration Using Tableau Software", Teimourzadeh and Kakavand introduce the concept of big data and provide a guideline on the integration of data analytics in management and business education and provide a pedagogical framework of developing data translation through the workflow stages.

Speaking of big data, video games produce terabytes of data, and game studios are investing more and more in game analytics to parse that data, understand their players, and improve their games. In Chapter "Understanding Players and Play Through Game Analytics", Tan, Katchabaw, and Slogar discuss video game data and game analytics. Here, effective data translation is needed since an important step in game analytics is to present the results to shareholders who may not be as familiar with the technical details of data science. These authors present a methodology for game analytics that employs clustering and visualization, discussing how to translate the player representation results from such analyses in an interpretable manner.

Redmond, Foucambert, and Libersan, in Chapter "Language Corpora and Principal Components Analysis", offer a brief overview and highlight challenges specific to the use of statistics in the field of linguistics. They provide an example of how Principal Components Analysis, a data reduction strategy that can be useful when dealing with data that has many variables, can be used to analyze a dataset of authentic texts of different genres written by post-secondary students in Quebec. The analysis was transformed into teaching modules designed to help post-secondary students improve their writing skills and to provide post-secondary instructors with an empirically-based framework to teach and evaluate genre-specific writing skills in French.

Chapter "A Tutorial of Analyzing Accuracy in Conceptual Change" by Li highlights the importance of ensuring training future data translators to understand whether the underlying assumptions for a given modeling method are, in fact, valid for their analyses. Through an analysis of simulated accuracy data from a conceptual change context, Li contrasts the use of two methods to analyze binary (i.e., correct/incorrect) responses. The flaws of employing the commonly used Analysis of Variance (ANOVA) method in this context are illustrated along with the advantages of a logistic regression-based modeling approach that is appropriate given the underlying nature of the response data of interest, especially in studying the near-ceiling performance. The chapter provides a strong illustration on the role of instructing students to understand the assumptions that underly methods used when translating data.

In Chapter "Transforming Data on the Boundaries of Science and Policy: The Council of Canadian Academies' Rhetorical Repertoire", Falconer turns our attention to the intersection of science and policy to transform and repurpose data for government policymakers, describing knowledge brokers as social actors who are working on the boundaries of science, translating data in ways that it might shape policy. Falconer refers to the ways that people communicate a message in an attempt to persuade different audiences. The need for flexibility by data translators is made evident.

The last chapter in our edited collection is Chapter "A Conceptual Framework for Knowledge Exchange in a Wildland Fire Research and Practice Context" by McFayden, Johnston, Woolford, George, Boychuk, Johnston, Wotton, and Johnston. This chapter presents a general conceptual knowledge exchange framework and illustrates its application to support the development of application-oriented research outcomes to inform operational decisions made in wildland fire management. The chapter provides an illuminating example of how translation skills among students in a data analytics consulting course are fostered, using fire management as the disciplinary foci. By providing a framework for considering how active learning across differing levels of expertise and disciplinary orientation may be useful, it sets the stage for our concluding discussion.

Collectively, the set of edited chapters in this volume highlights communication and knowledge translation, using data, and across disciplines. Throughout these chapters, interdisciplinarity emerges as a key theme. Pedagogically, training and educating future data translators requires more integration across discipline experts and conventional data scientists in a setting that encourages true knowledge exchange. The interdisciplinarity of this type of training requires intentionality. Consequently, we conclude this edited book with a chapter that connects key themes and focuses on the pedagogical implications that can guide an integrated and interdisciplinary approach to data translation.

This edited volume is a unique and timely contribution to this bourgeoning demand for data translators. The authors all demonstrate extraordinary efforts to bridge disciplines in their science, education, and policy work. Indeed, the editors are representative of this necessary interdisciplinarity that is evident in this book. Woolford is a statistical scientist. Kotsopoulos' expertise is in pedagogy. Samuels' expertise is in writing. It was the intersecting perspectives of each of the three editors, grounded in a problem of data translation, that became the seed for this volume.

London, Canada   Douglas G. Woolford
London, Canada   Donna Kotsopoulos
Toronto, Canada   Boba Samuels

Acknowledgments

Woolford, Kotsopolous, and Samuels are supported in part by funding from the Social Sciences and Humanities Research Council of Canada. We express our sincere gratitude to the esteemed contributing authors. This edited volume undertook a rigorous review process with each chapter receiving up to five reviews, both by the editorial team and external reviewers. Consequently, we are also grateful to the collection of experts who provided external reviews that guided the editors' reviews and the revisions that followed. Finally, we would also like to express our collective deep appreciation to our research assistant and graduate student, Brandon A. Dickson, who provided exemplary support to this project and our authors, demonstrating the true value of interdisciplinarity.

Contents

Translating Science into Actionable Policy Information—A Perspective on the Intergovernmental Panel on Climate Change Process

Francis W. Zwiers and Xuebin Zhang

Abstract Over its roughly 35-year history and six climate change assessments, the Intergovernmental Panel on Climate Change (IPCC) has evolved a set of calibrated terms that it uses to describe certainty and uncertainty in our understanding of the various facets of climate science. The careful and consistent use of this calibrated language over several assessment cycles has allowed the users of its reports to follow the progression of our understanding of climate change, its implications, and options for its mitigation. This chapter describes the development of the IPCC calibrated language and discusses its relation to statistical concepts of uncertainty. The chapter illustrates the application of the IPCC's calibrated language by describing how the level of confidence in the understanding of the causes of the observed global warming has evolved over time.

Keywords Intergovernmental Panel on Climate Change · Climate change · Climate models

The in-depth assessments of the science of climate change, its impacts, and adaption and mitigation options that are performed periodically by the Intergovernmental Panel on Climate Change (IPCC) can be viewed as an endpoint of an intense and wide-ranging body of scientific activity. In the case of the IPCC, this includes consideration of theory, evidence derived from hundreds of terabytes of observed climate and weather data, and from tens of petabytes of climate model output. The IPCC is charged with synthesizing policy-relevant information from thousands of published analyses of all of this data and information (~14,000 peer-reviewed papers, in the case of the most recent report from IPCC Working Group 1; WG1 [12]). The ultimate results of this activity are concise summaries for policymakers that contain the understanding of the science as agreed by the governments participating in the

F. W. Zwiers (✉)
Pacific Climate Impacts Consortium, University of Victoria, Victoria, BC, Canada
e-mail: fwzwiers@uvic.ca

X. Zhang
Climate Research Division, Environment and Climate Change Canada, Toronto, ON, Canada
e-mail: Xuebin.Zhang@ec.gc.ca

IPCC process. The last point is important—the IPCC summaries for policymakers are negotiated documents. This is a necessary step to ensure that the information that is subsequently used to negotiate international climate action via the United Nations Framework Convention on Climate Change (UNFCCC) does not, itself, become a point of negotiation at the periodic UNFCCC Conferences of the Parties, such as the COP-26 meeting held in Glasgow in November 2021.

The IPCC, which was established in 1988, issued its first assessment report in 1990 [2, 3] and recently completed three new comprehensive assessment reports, including that which deals with the physical science (WG1 [12]) as the main part of its work for its sixth assessment cycle. In addition, it typically produces between one and three special reports with a more topical focus during each roughly seven-year assessment cycle. IPCC assessments, which span a broad range of disciplines and involve thousands of experts, form the essential bridge between our understanding of climate change and its impacts on the one hand, and on the other hand, the actions that can be taken to limit change.

A key requisite for using science to develop policy and support international climate negotiations is to be able to communicate climate change assessments in a way that makes assessments comparable from one report to the next. This is required so that policy and decision makers can understand how expert consensus on climate change and response options has evolved over time. The IPCC has therefore developed, over successive assessment cycles, an approach to the description of certainties and uncertainties that is applicable across the breadth of the IPCC assessment, can be understood by scientists and users alike, and that enables the evaluation of changes in confidence and the strength of evidence from one assessment to the next.

The IPCC fifth and sixth assessment cycles use a three-tier lexicon ([5], hereafter M+2010) for describing certainty and uncertainty in our understanding of the various facets of climate science. This lexicon provides a framework that accommodates the different approaches that had previously evolved in the IPCC's three working groups, which deal with the physical science basis (Working Group I; WG1), climate change impacts, adaptation and vulnerability (Working Group 2; WG2), and climate change mitigation (Working Group 3; WG3).

Early WG1 and WG3 reports of the IPCC [2–4] did not use calibrated language to qualify findings, although WG1's [8] assessment that

> The balance of evidence suggests that there is a discernible human influence on climate [8, p. 4]

Portends the likelihood assessments that would become a feature of later WG1 reports. This statement hinted that it was more likely than not that some part of the change in our climate from pre-industrial conditions that had been observed by the early 1990s was due to human-induced greenhouse gas increases. In contrast to WG1, WG2 [13] did use some simple calibrated language to describe the confidence in its key findings, basically linking levels of confidence (i.e., *low, medium,* and *high*) to the quantity and consistency of the evidence supporting individual assessments.

WG2 [13] recognized that the process of assessing confidence levels was subjective, and that different individuals would likely assign confidence levels differently.

Moss and Schneider ([7], hereafter MS2000) therefore developed a set of calibrated terms that could be used to characterize uncertainty in a consistent way so that subjective variation in assessments could be reduced, and to enable comparison of assessments over time as the understanding of climate change evolved. They set out a detailed assessment procedure, demonstrated its application in a number of examples, and urged the use of a quantitative confidence scale, arguing that a numerical scale would help avoid the fact that uncertainty words have different meanings for different people. They thus suggested that the IPCC use five confidence levels: *very low* (0–0.05), *low* (0.05–0.33), *medium* (0.33–0.67), *high* (0.67–0.95) and *very high* (0.95–1.0), being careful to point out that these numerical confidence values should not be confused with the statistical notion of confidence. They also suggested a supplementary approach to uncertainty characterization based on the amount of evidence (e.g., observations, model output, theory, etc.) and its level of agreement/consensus, using four terms:

- *Speculative*—plausible ideas, but low evidence and low agreement
- *Competing explanations*—much evidence but low agreement
- *Established but incomplete*—high agreement, but not yet sufficient evidence
- *Well-established*—high agreement across much evidence

The MS2000 proposals were adopted in 2001 WG1 [9] and WG2 [14] reports, albeit with variations between the two working groups. In contrast, the WG3 [16] report did not use calibrated language in its assessments.

WG2 [14] made assessments as proposed by MS2000. In contrast, WG1 [9] used a seven-level scale rather than the five-level scale proposed by MS2000 to express judgments of confidence, and gave the scale a probabilistic interpretation as subjectively determined probabilities that the statements to which they were attached were true. The scale used in WG1 [9] has the following terms: *virtually certain* (greater than 99% chance that a statement is true); *very likely* (90–99% chance); *likely* (66–90% chance); *medium likelihood* (33–66% chance); *unlikely* (10–33% chance); *very unlikely* (1–10% chance) and *exceptionally unlikely* (less than 1% chance). As a key example of the application of this scale, WG1 [9] concluded that

> In the light of new evidence and taking into account the remaining uncertainties, most of the observed warming over the last 50 years is *likely* to have been due to the increase in greenhouse gas concentrations (p. 10).

The use of calibrated language became increasingly pervasive in the subsequent IPCC reports. In its fourth assessment report, WG1 [10] modified the likelihood scale slightly by clarifying that a *likely* assessment, for example, would not exclude the possibility that the statement could *very likely* be true. Thus, the scale was now given as *virtually certain* (>99% probability), *very likely* (>90%), *likely* (>66%), *about as likely as not* (33–66%), *unlikely* (<33%), *very unlikely* (<10%), and *exceptionally unlikely* (<1%). Three other levels were also added that were used in some cases: *extremely likely* (>95%), *more likely than not* (>50%), and *extremely unlikely* (<5%). WG1 [10] strengthened the previous assessment of the causes of the observed warming by saying that

Most of the observed increase in global average temperatures since the mid-20th century is *very likely* due to the observed increase in anthropogenic greenhouse gas concentrations (p. 10).

WG1 [10] also used a five-level confidence scale to assess statements of scientific understanding of a more general nature than the more specific outcomes that were assessed on the likelihood scale. The WG1 [10] confidence scale was described in units of "chance", with confidence terms and levels described as *very high* (at least 9 out of 10 chances), *high* (about 8 chances), *medium* (about 5 chances), *low* (about 2 chances) and *very low* (less than 1 out of 10 chances), with *low* and *very low* to be used only for major areas of concern. As it turned out, IPCC WG1 authors found this combination of two numerical scales difficult to use, with the result that confidence assessments were not frequently made, although confidence language does appear in some key assessments.

WG2 [15] similarly adopted a combination of likelihood and confidence scales for their assessments, using the same confidence scale as WG1 [10] and the seven-level likelihood scale that was used in WG1 [9]. They also occasionally used the three-level confidence scale of earlier WG2 reports. In contrast, WG3 [17] used a refinement of the evidence/agreement approach proposed by MS2000 to characterize its confidence in its assessments. Specifically, they evaluated whether there was *limited*, *medium* or *much evidence* to form the basis of an assessment and whether there was *low*, *medium* or *high* agreement amongst the available evidence. These evidence and agreement assessments were reported directly. As an example, a key WG3 [17] assessment that remains relevant today is that

In order to stabilize the concentration of GHGs in the atmosphere, emissions would need to peak and decline thereafter. The lower the stabilization level, the more quickly this peak and decline would need to occur. Mitigation efforts over the next two to three decades will have a large impact on opportunities to achieve lower stabilization levels (*high agreement, much evidence*) (p. 15).

WG3 took the evidence/agreement approach because of the nature of its remit, which includes the evaluation of technologies and policies for climate change mitigation.

M+2010 developed a tiered framework that unifies the approaches for making calibrated assessments across the three working groups. It did this by articulating a logic by which the likelihood, confidence and evidence/agreement approaches used predominantly by WG1, WG2 and WG3 respectively, can be related to each other. This framework uses evidence/agreement assessments as the basis for confidence assessments, which, in turn, may form the basis for likelihood assessments. The evidence/agreement and confidence assessments both use non-numerical scales that are delimited with defined qualifiers. In contrast, likelihood assessments are made when there is sufficient confidence in the available evidence and when the nature of the evidence makes a quantified uncertainty assessment possible. IPCC authors were advised that the available evidence should be assessed as *limited*, *medium* or *robust* depending on its type, amount, quality and consistency, and that agreement across the available evidence should be assessed as *low*, *medium* or *high*. They were also advised that confidence in a statement synthesized from the evidence should be expressed on a

High agreement Limited evidence	High agreement Medium evidence	High agreement Robust evidence
Medium agreement Limited evidence	Medium agreement Medium evidence	Medium agreement Robust evidence
Low agreement Limited evidence	Low agreement Medium evidence	Low agreement Robust evidence

Agreement ↑

Evidence (type, amount, quality, consistency) ⟶

Confidence Scale

Fig. 1 A depiction of evidence and agreement statements and their relationship to confidence. Confidence increases towards the top-right corner, as suggested by the increasing strength of shading. Generally, evidence is most robust when there are multiple, consistent independent lines of high-quality evidence. Figure from M+2010

five-level scale—*very low, low, medium, high* and *very high*—with confidence being lowest when there is limited evidence with low agreement, and highest when there is robust evidence with high agreement. The relationship between evidence/agreement assessments and confidence assessments is shown in Fig. 1. It was suggested that likelihood assessments could be made if confidence in a statement was high, or very high and a quantified probabilistic uncertainty assessment was possible. Likelihood assessments could also occasionally be made if there was only medium confidence in the evidence provided the confidence assessment was explicitly reported.

The assessments of the causes of the observed warming since pre-industrial times using the M+2010 guidance provide an example of how the IPCC has continued to produce assessments whose evolution can be followed over time. WG1 [11] stated in its 5th assessment report that

> Human influence has been detected in warming of the atmosphere and the ocean, in changes in the global water cycle, in reductions in snow and ice, in global mean sea level rise, and in changes in some climate extremes (…). This evidence for human influence has grown since AR4. It is *extremely likely* that human influence has been the dominant cause of the observed warming since the mid-20th century (p. 17).

Based on broad and diverse lines of evidence, this assessment assigns a higher likelihood to the statement than previous assessments of the human influence on climate. It also indicates a more confident assessment of the magnitude of the human-induced warming, replacing the word "most" that was used in WG1 [9] and WG1 [10], meaning at least half of the observed warming, with the word "dominant". This statement was supported with an additional assessment that "It is *extremely likely* that more than half of the observed increase in global average surface temperature from 1951 to 2010 was caused by [human influence]" (p. 17).

WG1 [12] has recently strengthened the assessment of the causes of warming even further by stating that:

It is unequivocal that human influence has warmed the atmosphere, ocean and land. Widespread and rapid changes in the atmosphere, ocean, cryosphere and biosphere have occurred (p. 4).

WG1 [12] attributes essentially all of the observed warming since the 1850–1900 period to human influence on the climate system, with the statement

The *likely* range of total human-caused global surface temperature increase from 1850–1900 to 2010–2019 is 0.8°C to 1.3°C, with a best estimate of 1.07°C (p. 5).

In the first of these two statements, the finding that human influence has warmed the climate is expressed as a fact not requiring qualification by attaching a likelihood to the statement—supporting the view that there is no remaining doubt of its truth. Instead, the focus of uncertainty quantification has shifted to the quantification of the magnitude of the human influence, which is given in the second statement as a Bayesian credible interval.

The confidence and likelihood terms used by the IPCC are reminiscent of terms used in statistics, but they should not be confused with the statistical notions of confidence and likelihood. IPCC likelihood assessments are often based on the results of multiple statistical studies, but the translation of those results, which are predominantly obtained from the frequentist perspective, into likelihood statements that have a Bayesian flavor requires expert judgment. While the approach is far from perfect, it has enabled consensus building on our scientific understanding of climate change and has been relatively consistently applied, with some modification, through the last four IPCC assessment cycles. Policymakers have therefore been able to follow the progression of IPCC's key assessments, such as those concerning the causes of global warming, over the 25-year period since the first assessment indicating a discernable human influence on climate was made [8] and can have confidence that those assessments have been made in a relatively consistent manner over time.

While the structured approach of M+2010 for making uncertainty (and certainty) assessments in the IPCC has been largely successful, IPCC authors still often find its application difficult. The application of the idealized relationship between evidence/agreement and confidence assessments is not always as straightforward as depicted in Fig. 1. For example, there are situations where, despite there being only a very limited number of studies, confidence may nevertheless be *very high* because there is no question concerning their robustness. In other cases, there may be much evidence that points in the same direction, but confidence is reduced because of concern about dependence that arises due to the common use of data and models. The determination of likelihood assessments can also be challenging, even when confidence is *very high*.

For example, the detection and attribution studies that are the basis for the assessments of the causes of global warming generally proceed by regressing observations onto signals of change that are extracted from climate model simulations driven with known changes in greenhouse gas concentrations, atmospheric aerosol loadings, land use, volcanic activity and solar output (e.g., see [1]). The results of these studies are (a) inferences about whether the climate model simulated signals can be found in observations, obtained via a hypothesis testing paradigm, (b) interval estimates of the scaling factors that are needed to adjust the amplitudes of the signals to best match observations, and (c) interval estimates of the amount of change that can be attributed to individual combinations of forcing agents. With a few exceptions, these inferences have been made with frequentist methods. Often, a number of such studies will be available using related but differently compiled observational datasets, a range of variations on the methods used for processing the data and performing the regression, and different choices of climate models for signal estimation and for estimating the characteristics of the natural, unforced variability of the climate system. The assessment challenge in such a situation is then to synthesize information about attributed changes from several studies, each providing imperfectly estimated confidence intervals, into a statement that reads like a Bayesian credible interval and has the form "[it is] *likely* [that the] total human-caused global surface temperature increase from 1850–1900 to 2010–2019 is 0.8 °C to 1.3 °C" [12, p. 5]. This is a non-trivial task that ultimately relies on expert judgment.

In the IPCC, the assessment of what likelihood to associate with a given interval estimate, such as the one above, involves consensus building amongst experts rather than direct interpretation of interval estimates. This process is made more difficult by confusion about the differences of interpretation between frequentist confidence intervals and Bayesian credible intervals, which are generally not well understood by IPCC authors. In the frequentist context, a $(1 - p) \times 100\%$ confidence interval is constructed so that it will include the unknown parameter of interest $(1 - p) \times 100\%$ of the time in repeated calculations under identical sampling conditions provided all assumptions that go into the calculation are satisfied. This provides indirect information about the parameter. In contrast, a Bayesian credible interval, which is a statement of the posterior probability (i.e., calculated conditional on the observational and modeling data that were considered) that the parameter of interest lies in the interval, provides a more direct inference about the parameter. The attributed warming statement highlighted above clearly has such a Bayesian interpretation, but it has been synthesized by expert judgment from a collection of climate change detection and attribution studies that used frequentist methods. Thus, while the statement does provide a quantified likelihood assessment in the sense intended by M+2010, it is fundamentally a product of expert judgment.

A second type of challenge that IPCC authors often face concerns the trade-off between specificity and confidence that must be made in almost every assessment that uses calibrated language. One example of this trade-off that particularly affects assessments at the regional scale is that we can often be more confident in the observed or projected *direction* of change than in the *magnitude* of change. This creates a quandary for activities such as regional adaptation planning that require confidence in the magnitude of change that is to be taken into account. Another example where such a trade-off occurs relates to spatial scale of assessments, as is illustrated by the assessment of changes in the intensity of extreme rainfall events. Many policymakers are preoccupied by the impacts of global warming on extreme rainfall because of the implications that it has for changes in the frequency of flash flooding, the design of infrastructure, and impacts to agriculture and ecosystems. By considering the global land area with sufficient data from weather stations to estimate long term trends in extreme precipitation accumulations, WG1 [12] was able to make the assessment that

> The frequency and intensity of heavy precipitation events have increased since the 1950s over most land area[s] for which observational data are sufficient for trend analysis (*high confidence*), and [that] human-induced climate change is *likely* the main driver (p. 8).

This relatively strong statement about the global scale intensification of extreme precipitation is based on a body of evidence that has been developing and strengthening steadily since evidence of human influence on precipitation [18] and its extremes [6] was first published, with dozens of studies now available. Nevertheless, the ability to make assessments about changes in extreme precipitation at the local and regional scales that are needed to trigger adaptation planning remains rather limited. This is illustrated by the WG1 [12] synthesis of regional changes in extreme precipitation that is shown in Fig. 2. As can be seen, there are many regions with insufficient data or that have not been sufficiently studied to make an assessment of whether human influence on the climate has affected precipitation extremes. In addition, there are a number of regions where studies are available, but agreement in the available evidence and analyses is lacking. Even in areas where the body of evidence does indicate an intensification of extreme precipitation, there is generally low confidence that this can be attributed to human influence.

The main reasons why the regional assessments of changes in extreme precipitation are much more clouded relate to (a) the large influence of natural internal climate variability on extreme precipitation, which tends to obscure the effects of anthropogenic forcing at regional and local scales, (b) the role that changes atmospheric circulation patterns may have, which remains poorly understood, and (c) regional variations in anthropogenic forcing that are not well represented in global climate models. Global-scale analyses more effectively filter out these different sources of "noise" to reveal the underlying impact of human influence than regional analyses. Thus, it is clear that there is a complex trade-off between the confidence that one can have in a particular assessment, the scale on which it is made and the specificity with which it is made. This is an aspect of the distillation of science that tends to be reasonably well understood by climate scientists, but is often substantially less well appreciated by policymakers and the non-scientists involved in negotiating the IPCC summaries for policymakers, who sometimes seek to over-interpret findings for the countries and regions they represent. The careful and consistent use of calibrated language in assessments can help to counter this kind of pressure.

In summary, the IPCC has successfully applied calibrated uncertainty language in the process of making policy-relevant assessments on a key societal issue. The use of this language (M+2010), which has evolved only slightly over several IPCC assessment cycles spanning about two and a half decades and is now used consistently throughout the IPCC, has allowed the IPCC to make key assessments that are comparable over time. Therefore, policymakers have received a reliable evaluation of the increasing strength of the evidence that human activity is affecting the climate in important ways. The use of this language also allows scientists to convey how

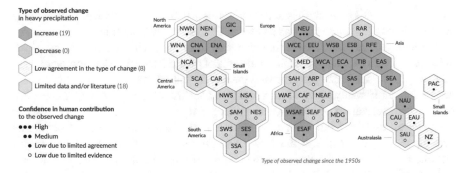

Fig. 2 Synthesis of assessed observed and attributable regional changes. [Displayed here is Panel (b) of Figure SPM.3 from WG1 [12], which reports assessment results for heavy precipitation.] The IPCC AR6 WGI inhabited regions are displayed as hexagons[1] with identical size in their approximate geographical location (see legend for regional acronyms). All assessments are made for each region as a whole and for the 1950s to the present. Assessments made on different time scales or more local spatial scales might differ from what is shown in the figure. The colours in each panel represent the four outcomes of the assessment on observed changes. Striped hexagons (white and light-grey) are used where there is *low agreement* in the type of change for the region as a whole, and grey hexagons are used when there is limited data and/ or literature that prevents an assessment of the region as a whole. Other colours indicate at least *medium confidence* in the observed change. The confidence level for the human influence on these observed changes is based on assessing trend detection and attribution and event attribution literature, and it is indicated by the number of dots: three dots for *high confidence,* two dots for *medium confidence* and one dot for *low confidence* (single, filled dot: limited agreement; single, empty dot: limited evidence). For heavy precipitation, the evidence is mostly drawn from changes in indices based on one-day or five-day precipitation amounts using global and regional studies. Green hexagons indicate regions where there is at least *medium confidence* in an observed increase in heavy precipitation

the degree of confidence that they have in their findings declines as the requests for assessment from policymakers become more specific. Nevertheless, there are significant challenges in the application of the uncertainty language, including those of

---

[1] Each hexagon corresponds to one of the IPCC AR6 WG1 reference regions: North America: NWN (North-Western North America, NEN (North-Eastern North America), WNA (Western North America), CNA (Central North America), ENA (Eastern North America), Central America: NCA (Northern Central America), SCA (Southern Central America), CAR (Caribbean), South America: NWS (North-Western South America), NSA (Northern South America), NES (North-Eastern South America), SAM (South American Monsoon), SWS (South-Western South America), SES (South-Eastern South America), SSA (Southern South America), Europe: GIC (Greenland/Iceland), NEU (Northern Europe), WCE (Western and Central Europe), EEU (Eastern Europe), MED (Mediterranean), Africa: MED (Mediterranean), SAH (Sahara), WAF (Western Africa), CAF (Central Africa), NEAF (North Eastern Africa), SEAF (South Eastern Africa), WSAF (West Southern Africa), ESAF (East Southern Africa), MDG (Madagascar), Asia: RAR (Russian Arctic), WSB (West Siberia), ESB (East Siberia), RFE (Russian Far East), WCA (West Central Asia), ECA (East Central Asia), TIB (Tibetan Plateau), EAS (East Asia), ARP (Arabian Peninsula), SAS (South Asia), SEA (South East Asia), Australasia: NAU (Northern Australia), CAU (Central Australia), EAU (Eastern Australia), SAU (Southern Australia), NZ (New Zealand), Small Islands: CAR (Caribbean), PAC (Pacific Small Islands).

interpreting results from studies that dominantly use frequentist statistical paradigms with calibrated language that has a natural, Bayesian interpretation.

We hope that this short chapter has provided some insight into the process by which science, through the analysis of vast amounts of observed and model-simulated data, can play a role in the negotiation and establishment of international policy. There are undoubtedly many areas of international concern where a systematic approach to the characterization of certainty and uncertainty can be helpful. Climate change is, however, distinct from other issues, such as the future emergence of pandemic causing pathogens, in that it has both wide ranging, potentially existential, implications for humans and a known cause—the rapid rise in atmospheric greenhouse concentrations that are the result of the human use of fossil fuel. The existence of this single cause, which was not yet irrefutably confirmed when the IPCC was created in 1988, led to a unique international assessment process that has four key characteristics—(i) the assessments are commissioned by the IPCC's member governments to evaluate the entire scientific literature on a broad collection of climate change related topics, the choice of which varies somewhat from one assessment cycle to the next; (ii) the process used to review the assessments, which solicits comments and criticism from the entire international scientific and policymaker community; (iii) the remit to perform an assessment as opposed to a review, making judgments about what is known with confidence, where consensus has been established and where it still remains to be achieved; and (iv) the requirement to defend the assessments and negotiate how they are communicated in an international, intergovernmental forum. The IPCC's calibrated uncertainty language has contributed substantially to the success of the process by providing governments with a means to track the evolution of our understanding of climate change, its implications and solutions, and by providing scientists with an important tool for communicating what is known and what remains uncertain in a consistent manner across multiple assessment cycles.

References

1. Eyring, V., Gillett, N.P., Achuta Rao, K.M., Barimalala, R., Barreiro Parrillo, M., Bellouin, N., Cassou, C., Durack, P. J., Kosaka, Y., McGregor, S., Min, S., Morgenstern, O., Sun, Y.: Human influence on the climate system. In: Masson-Delmotte, V., Zhai, P., Pirani, A., Connors, S.L., Péan, C., Berger, S., Caud, N., Chen, Y., Goldfarb, L., Gomis, M.I., Huang, M., Leitzell, K., Lonnoy, E., Matthews, J.B.R., Maycock, T.K., Waterfield, T., Yelekçi, O., Yu, R., Zhou, B. (eds.) Climate Change 2021: The Physical Science Basis. Contribution of Working Group I to the Sixth Assessment Report of the Intergovernmental Panel on Climate Change. Cambridge University Press (2021) (In Press). https://www.ipcc.ch/report/sixth-assessment-report-working-group-i/
2. IPCC: In: Houghton, J.T., Jenkins, G.J., Ephraums, J.J. (eds.) Climate Change: The IPCC Scientific Assessment. Cambridge University Press (1990). https://www.ipcc.ch/report/climate-cha nge-the-ipcc-1990-and-1992-assessments/
3. IPCC: Climate Change: The IPCC 1990 and 1992 Assessments. Intergovernmental Panel on Climate Change (1992). ISBN 0-662-19821-2. https://www.ipcc.ch/report/climate-change-the-ipcc-1990-and-1992-assessments/

4. IPCC: Climate Change 1995: The IPCC Second Assessment. Intergovernmental Panel on Climate Change (1996). https://www.ipcc.ch/site/assets/uploads/2018/05/2nd-assessment-en-1.pdf
5. Mastrandrea, M.D., Field, C.B., Stocker, T.F., Edenhofer, O., Ebi, K.L., Frame, D.J., Held, H., Kriegler, E., Mach, K.J., Matschoss, P.R., Plattner, G.-K., Yohe, G.W., Zwiers, F.W.: Guidance Note for Lead Authors of the IPCC Fifth Assessment Report on Consistent Treatment of Uncertainties. Intergovernmental Panel on Climate Change (2010). https://www.ipcc.ch/site/assets/uploads/2018/05/uncertainty-guidance-note.pdf
6. Min, S.-K., Zhang, X., Zwiers, F.W., Hegerl, G.C.: Human contribution to more intense precipitation events. Nature **470**, 378–381 (2011). https://doi.org/10.1038/nature09763
7. Moss, R.H., Schneider, S.H.: Uncertainties in the IPCC TAR: Recommendations to lead authors for more consistent assessment and reporting. In: Pachauri, R., Taniguchi, T., Tanaka, K. (eds.) Guidance Papers on the Cross Cutting Issues of the Third Assessment Report of the IPCC. Intergovernmental Panel on Climate Change (2000). ISBN 4-9980908-0-1. https://digitallibrary.un.org/record/466950?ln=en
8. WG1: In: Houghton, J.T, Meira Filho, L.G., Callander, B.A., Harris, N., Kattenberg, A., Maskell, K. (eds.) Climate Change 1995: The Science of Climate Change. Contribution of Working Group I to the Second Assessment Report of the Intergovernmental Panel on Climate Change. Cambridge University Press (1996). https://www.ipcc.ch/report/ar2/wg1/
9. WG1: In: Houghton, J.T., Ding, Y., Griggs, D.J., Noguer, M., van der Linden, P.J., Dai, X., Maskell, K., Johnson, C.A. (eds.) Climate Change 2001: The Scientific Basis. Contribution of Working Group I to the Third Assessment Report of the Intergovernmental Panel on Climate Change. Cambridge University Press (2001). https://www.ipcc.ch/report/ar3/wg1/
10. WG1: In: Solomon, S., Qin, D., Manning, M., Chen, Z., Marquis, M., Averyt, K.B., Tignor, M., Miller H.L. (eds.) Climate Change 2007: The Physical Science Basis. Contribution of Working Group I to the Fourth Assessment Report of the Intergovernmental Panel on Climate Change. Cambridge University Press (2007). https://www.ipcc.ch/report/ar4/wg1/
11. WG1: In: Stocker, T.F., Qin, D., Plattner, G.-K., Tignor, M., Allen, S.K., Boschung, J., Nauels, A., Xia, Y., Bex, V., Midgley, P.M. (eds.) Climate Change 2013: The Physical Science Basis. Contribution of Working Group I to the Fifth Assessment Report of the Intergovernmental Panel on Climate Change. Cambridge University Press (2013). https://www.ipcc.ch/report/ar5/wg1/
12. WG1: In: Masson-Delmotte, V., Zhai, P., Pirani, A., Connors, S.L., Péan, C., Berger, S., Caud, N., Chen, Y., Goldfarb, L., Gomis, M.I., Huang, M., Leitzell, K., Lonnoy, E., Matthews, J.B.R., Maycock, T.K., Waterfield, T., Yelekçi, O., Yu, R., Zhou, B. (eds.) Climate Change 2021: The Physical Science Basis. Contribution of Working Group I to the Sixth Assessment Report of the Intergovernmental Panel on Climate Change. Cambridge University Press (2021) (In Press). https://www.ipcc.ch/report/sixth-assessment-report-working-group-i/
13. WG2: In: Watson, R.T., Zinyowera, M.C., Moss, R.H. (eds.) Climate Change 1995: Impacts, Adaptations and Mitigation of Climate Change. Contribution of Working Group II to the Second Assessment Report of the Intergovernmental Panel on Climate Change. Cambridge University Press (1996). https://www.ipcc.ch/report/ar2/wg2/
14. WG2: In: McCarthy, J.J., Canziani, O.F., Leary, N.A., Dokken, D.J., White, K.S. (eds.) Climate Change 2001: Impacts, Adaptation and Vulnerability. Contribution of Working Group II to the Third Assessment Report of the Intergovernmental Panel on Climate Change, Cambridge University Press (2001). https://www.ipcc.ch/report/ar3/wg2/
15. WG2: Climate Change 2007: Impacts, Adaptation and Vulnerability. Contribution of Working Group II to the Fourth Assessment Report of the Intergovernmental Panel on Climate Change. Cambridge University Press (2007). https://www.ipcc.ch/report/ar4/wg2/
16. WG3: In: Davidson, O., Metz, B. (eds.) Climate Change 2001: Mitigation. Contribution of Working Group II to the Third Assessment Report of the Intergovernmental Panel on Climate Change. Cambridge University Press (2001). https://www.ipcc.ch/report/ar3/wg3/

17. WG3: In: Metz, B., Davidson, O.R., Bosch, P.R., Dave, R., Meyer, L.A. (eds.) Climate Change 2007: Mitigation. Contribution of Working Group III to the Fourth Assessment Report of the Inter-governmental Panel on Climate Change. Cambridge University Press (2007). https://www.ipcc.ch/report/ar4/wg3/
18. Zhang, X., Zwiers, F.W., Hegerl, G.C., Lambert, F.H., Gillett, N.P., Solomon, S., Stott, P.A., Nozawa, T.: Detection of human influence on twentieth-century precipitation trends. Nature **448**, 461–465 (2007). https://doi.org/10.1038/nature06025

Data in Observational Astronomy

Pauline Barmby and Samantha Wong⊙

Abstract Astronomy is arguably the first data science. Astronomical observations date back to prehistoric times: early peoples used observations of the Sun, Moon, stars, and planets for navigation, timekeeping, and many other purposes. Ancient cultures catalogued the position and brightness of stars and planets. Throughout history, the sky has been of interest to scientists and non-scientists alike. The data that astronomers use to make discoveries are both the lifeblood of the discipline and a source of wonder and inspiration. This chapter provides an introduction for the non-specialist, describing why astronomers collect data; how data are collected, processed, used and shared, both between astronomers and between astronomers and the public; unique aspects of astronomical data; and future challenges for telling the story of the universe with astronomical data.

Keywords Astronomical data · Astronomical telescopes · Astronomical imaging · Astronomical spectroscopy · Astronomical catalogues

1 Why Do Astronomers Collect Data?

Researchers in astronomy use observational data to describe and measure the contents of the universe, test physical theories, and search for new phenomena. Astronomical observations are driven by both technology development and astrophysical theory. New observatories and technologies enable observational data to be collected with improved resolution in space, time, and energy. The development and application of astrophysical theory enable new predictions about the formation, evolution, and operation of the contents of the universe that must be tested by comparison with empirical observations.

P. Barmby (✉)
Department of Physics and Astronomy and Institute for Earth and Space Exploration, Western University, London, ON, Canada
e-mail: pbarmby@uwo.ca

S. Wong
McGill University, Montreal, QC, Canada
e-mail: swong@physics.mcgill.ca

The universe is not static. Objects in the Earth's vicinity—planets, comets and asteroids in the solar system, stars in the Sun's neighbourhood of the Milky Way— show measurable sky motion relative to more distant objects. Those motions can be used to understand the objects' history and provenance as well as their relationship to other nearby phenomena. The brightness of celestial objects also changes. Stars vary in brightness for many reasons, including the effects of interior physical processes or interactions with a companion star. Stars and stellar objects such as white dwarfs and neutron stars can suffer catastrophic explosions. Even with a perfect view of the contents of today's universe, the need to collect astronomical data would continue.

2 What Are Astronomy Data?

Astronomical data come in multiple forms. The least processed are the raw data produced by sensors and detectors attached to telescopes and other instruments. Today, such data are almost always in digital format. Historical astronomical data can also be in the form of photographs on film or plates, or records of human visual observations. Since photographic media have limited lifetimes, the archival collections of many observatories are being digitized [18]. The raw data produced at observatories is usually processed and calibrated so that it can be used to measure physical properties of celestial objects on a quantitative scale. The resulting measurements are often combined into catalogues—datasets that include multiple properties of many sources. Objects can be selected to comprise a catalogue for many reasons: for example, a catalogue could include all objects emitting detectable ultraviolet radiation in a specific area of sky, or all the known stars of a specific type in the Sun's neighbourhood of the Milky Way, or all the galaxies in a galaxy cluster.

The digital data from observatories are often preserved for their long-term archival value. Some facilities produce raw data at extremely high rates, and preserving all data is not always possible; this will be even more true with near-future facilities such as the Square Kilometer Array radio telescope [7]. As described above for data obtained with photographic technology, changes in data acquisition or data storage technology can threaten the long-term viability of archives. Maintaining access to historical astronomical data is an ongoing challenge.

2.1 Where Do Astronomy Data Come from?

This chapter focuses on observational data obtained via the electromagnetic radiation received from astronomical sources; a more technical introduction to the same topic is given by Barmby [3].[1] The term electromagnetic (EM) radiation describes the

[1] Not included here are observational data acquired using other 'messengers' from the cosmos including gravitational waves and elementary particles such as neutrinos (see, e.g., [25]), or the

energy that is transported in the form of either photons or electromagnetic waves [43]. Physicists and astronomers think of light either as waves of electric and magnetic fields travelling through space, or countable particles called photons. Because light can be characterized as a wave, electromagnetic radiation is often described by wave properties, such as wavelength (the distance between successive peaks or troughs), frequency (the number of oscillations within a given time), and amplitude, which gives the amount of energy contained in the wave.

EM radiation forms the electromagnetic spectrum, ranging from long-wavelength radio waves to short-wavelength, high-frequency gamma rays (Fig. 1). The human eye can detect only visible light, which comprises a tiny fraction of the entire EM spectrum. Observations with visible light were the first to be made in astronomy; today's electronic sensors record visible light with much higher efficiencies than the human eye. While many familiar objects such as stars, planets, and galaxies are detectable at visible wavelengths, many other astronomical objects and phenomena remain invisible to the human eye. Specialized sensors and detectors are needed to record EM radiation at, for example, ultraviolet and radio wavelengths.

The distribution across the electromagnetic spectrum of a celestial object's emitted radiation depends on the object's properties. Colours of stars indicate temperature, with hotter stars emitting bluer light than cooler stars. In general, hotter objects, such as young stars and the hot gas around black holes, emit light at the bluer end of the

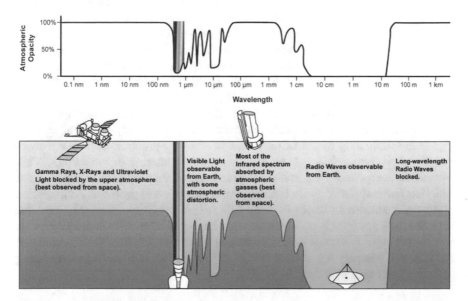

Fig. 1 The electromagnetic spectrum and its transmittance through Earth's atmosphere. *Credit* NASA, public domain via Wikimedia Commons, https://commons.wikimedia.org/wiki/File:Atmospheric_electromagnetic_opacity.svg

data produced by computer simulations of astrophysical processes, which can be quite extensive (e.g., 5 petabtyes for the simulation described by Heitmann et al. [17]).

spectrum—that is, blue light to ultraviolet through high energy gamma rays. Infrared radiation has longer wavelengths than the red light of the visible spectrum and typically arises from planets, star forming regions, and clouds of interstellar dust and gas warmed by starlight. Radio waves range from metres to kilometres in wavelength. These are emitted from the coolest objects in the universe, including cold interstellar gas and dust as well as the "cosmic microwave background", leftover radiation from the early universe. Many celestial objects contain multiple components with different temperatures and combining observations made at different wavelengths can give information on an object's properties.

Astronomers collect electromagnetic radiation using both ground and space-based telescopes. The different types of EM radiation require different instruments for data collection, with instrument properties chosen to receive specific wavelengths. Because the Earth's atmosphere scatters and absorbs incoming photons, telescope locations can vary depending on the wavelength involved. Telescopes that observe high-frequency light (gamma rays, X-rays, ultraviolet radiation) need to be in space rather than on Earth's surface. For very high energy gamma-rays, small space-based detectors do not have the surface area required to capture enough photons. Fortunately, gamma-ray photons decay in the upper atmosphere into lower energy particles that travel faster than the speed of light in air and create optical Cherenkov radiation (see [39]) that can be captured by large ground-based optical telescopes (e.g., VERITAS; [40]). For visible and infrared wavelengths, ground-based observation is possible in distinct ranges of wavelengths that can penetrate the atmosphere (see Fig. 1). Although the atmosphere transmits visible and infrared light, it also distorts that light, resulting in images that are blurred. Space-based telescopes such as the Hubble Space Telescope and the James Webb Space Telescope provide an undistorted view but are much more expensive to build and operate than ground-based telescopes.

In most cases, astrophysics is an observational rather than an experimental science. Astrophysicists cannot directly manipulate the objects they study. Nearly all celestial objects are studied by means of the radiation that they produce, or their effects on radiation from other objects (e.g., via absorption, scattering or reflection). While astronomers can control the statistical properties of data samples to some extent, their data and sample sizes are limited to what the universe provides. Astronomical data samples are also limited by the availability of telescope time. Most telescopes can only observe a tiny fraction of the sky at once; acquiring larger samples requires more observation time. Telescope time is a precious resource, competitively allocated, and experimental samples are not always large enough to yield analyses with strong evidence for or against a specific hypothesis.

2.2 How Are Astronomy Data Processed?

The raw data collected from telescopes can be broadly characterized into two types: imaging and spectroscopy (Fig. 2). Imaging refers to the familiar two-dimensional

Fig. 2 Astronomical data
cube, showing imaging data
(brightness as a function of
sky position) along two axes
and spectroscopic data
(brightness as a function of
wavelength) on the third
axis. *Credit* DSI, University
of Stuttgart. Reproduced
with permission

picture in which a *camera* is used to collect photons originating from a specific region
of sky, usually in a carefully chosen range of wavelengths. In some wavelength
regimes, the arrival time of each photon is measured. Spectroscopy refers to the
distribution of object brightness as a function of wavelength, (usually over a relatively
small wavelength range) collected with a *spectrograph* and recorded digitally. Some
instruments combine these two methods, recording spectra at multiple sky positions
simultaneously and producing a data cube.

Both imaging and spectroscopic data are typically stored and transmitted in a
standard file format known as the Flexible Image Transportation System (FITS; [29]).
The FITS format has been the standard for astronomical data since the late 1970s [35].
FITS files have remained essentially unchanged since their introduction, consisting of
a three-dimensional data matrix with coordinate axes and flux values (representing
brightness measured at a particular position on the sky), as well as a header with
metadata in human-readable form that contains information about the observation.
The header format is relatively flexible, with a wide selection of optional keyword-
value pairs that can be accompanied by a plain language comment. Headers often
include information about the instrument used to record the data, the configuration
of the telescope, any post-observation processing that has been performed, and other
useful information.

While FITS is best-known as an image data format, it also includes a standard for
storing tabular data [29]. Both FITS tables and FITS images can contain checksums,
allowing data corruption to be detected. One common use case for FITS tables is
storage of astronomical catalogs comprising measurements of observed objects such
as sky location, brightness at different wavelengths, spatial extent, or classification.

Many astronomical catalogs are publicly available, allowing astronomers and non-astronomers alike to re-use this information for their own projects. Many catalogs are made publicly available via services such as VizieR [28], or incorporated into two widely used literature compilation databases, SIMBAD [42] and NED [23]. These databases are continuously updated as new objects are discovered or new observations determine more information about known objects. Individual astronomical surveys such as the *Gaia* mission [13] also maintain their own publicly available databases; Weijmans et al. [41] discuss the considerations that go into completing one of many public data releases from the Sloan Digital Sky Survey.

2.3   How Are Astronomy Data Visualized and Used?

Astronomers typically use specialized software (e.g., AstroImageJ, described by Collins et al. [8]; or ds9, described by Joye [20]) to visualize raw images and spectra and track data processing, because they are interested in the quantitative physical information that can be extracted from those data. For example, astronomers want to know the exact numerical value associated with a given pixel in an image, because it corresponds to a calibrated measurement of an object's brightness. They want to know the exact centre of the distribution of pixels that make up an image of a galaxy because that allows measurement of the absolute sky location of the galaxy. In processing images for scientific purposes, astronomers do not generally use image-processing software such as might be used in digital photography: the algorithms in such software can change the pixel values in unpredictable ways and alter the calibrated relationship between pixel value and physical brightness.[2]

Once processed and calibrated, astronomical data are used to estimate values of physical quantities, such as brightness of an object in an image, brightness of a feature in a spectrum, or position of an object on the sky. These measurements may be used to test models relating to a particular object (e.g., the predicted orbit of an asteroid determines its position on the sky) or to a class of objects (e.g., the probability distribution function that describes their brightnesses, or spatial positions). Exploratory data analysis might involve plotting a (hypothesized) independent variable against a prediction made from that variable and measured by the observations. Techniques for dimensionality reduction in exploratory data analysis, such as principal component analysis, self-organizing maps, and t-distributed stochastic neighbor embedding, are becoming widely used, especially for large catalog datasets; Baron [4] gives a concise introduction. The use of astronomical datasets to measure physical quantities means that estimating measurement uncertainty is extremely important. This can be done by repeated measurements of the same quantity from different raw data or by estimating

---

[2] Such software is often used in producing aesthetic images, for example as part of a press release. See Sect. 4.2 below.

the measurement noise from the calibrated data itself. The incorporation of measurement uncertainties in machine-learning algorithms is an area of active research in which astronomers and astro-statisticians are deeply involved (e.g., [15, 33]).

3 Unique Aspects of Astronomy Data

Most astronomical observational facilities are publicly funded. Since every observation is potentially unique and observational resources are scarce, many funding agencies mandate that astronomy data should be publicly available. In contrast with most sciences, a significant fraction of all astronomy data is public: anyone with Internet access can download FITS-formatted images and catalog data for their own research or study. Pössel [31] provides an extensive guide for new users of astronomical data. Astronomy has a strong tradition of citizen scientists—that is, volunteers without formal astronomy education—who can make use of public data to make discoveries of their own and contribute to ongoing astronomy research [22]. The degree to which data are processed before being made available depends on the observational facility. Historically, space astronomy missions have had the resources to provide processed data to users, while ground-based telescope data are often only minimally processed; this is changing with the advent of large ground-based survey facilities such as the Legacy Survey of Space and Time [19].

Most of the tools that astronomers use for analyzing their data are open-source and freely available, often in the form of community-developed packages (e.g., [38]). These tools can be downloaded and used by anyone and interested people with a programming background can contribute to the code without formal training in astronomy. Compared to other disciplines, astronomical data have few commercial or privacy restrictions. Astronomy can thus provide excellent datasets for the development of machine learning and data mining algorithms.

4 Sharing Astronomy Data and Results

4.1 Astronomy Data Sharing Between Researchers

The methods used by astronomers to share data with each other depend on the type of data being shared. As discussed above, raw and/or processed data are often made available by the observational facilities themselves. Following observation and processing, most astronomical data are made available in publicly accessible archives, sometimes with a short 'proprietary period' in which they are only accessible to the team that designed and proposed the corresponding observations. This mechanism incentivizes astronomers to develop efficient programs that make the best possible use of valuable facilities.

Publication in the professional literature is another type of data-sharing, and electronic publication of astronomical catalogues in the professional literature is very common. Projects which receive large allocations of telescope time often commit to making public releases of both raw and processed data; such releases can be from a team's own website or through a national facility such as the Canadian Astronomy Data Centre [14] or the US National Optical-Infrared Astronomy Research Laboratory Data Lab [26]. Third-party data-sharing sites such as Figshare or Zenodo are sometimes used by astronomers, and the field is still developing data citation practices [27, 30].

Since the advent of electronic telescope archives, astronomers have been working toward the ideal of the virtual observatory, a web-based assembly of data archives as well as analysis and exploration tools [10]. At the present day, there is not one "virtual observatory", but a series of national efforts coordinated by the International Virtual Observatory Alliance [5]. A complete 'one-stop shop' for all the data from all the world's telescopes does not yet exist. Despite standardized file formats and software, there are still substantial technical, financial, and logistical obstacles to combining data from multiple observatories and making it easily searchable [16].

4.2 Astronomy Data Sharing with the Public

Images are an important component of astronomical datasets. Their inherently visual nature provides an effective method for communicating astronomy data and current research data to the public. However, the FITS images studied by astronomers only show the intensity of the radiation received in a single waveband, usually represented in greyscale. Processing astronomical data into the colourful images seen in popular science magazines and observatory press releases involves both art and science. These images are not "true colour" in that they do not represent what an unaided human eye would see if close to the source. This is not a deliberate deception, but rather a consequence of the limitations of human vision compared to astronomical instruments.

The grayscale images used by astronomers are often converted to colour images by combining images made with red, green, and blue filters. The colours seen in these images are essentially what one would see if looking at a bright white light through the filter that was used in the telescope [1, 32]. Images made using non-visible wavelengths of light may also be processed to use false colour to display various elements of the object or phenomenon, such as motion, depth, and energy. The way in which colours and shading are used to represent these properties is known as visual grammar and serves to give two-dimensional images extra detail that can help the viewer without a scientific background understand the image [11, 32].

There are many strategies for effectively communicating the information contained within astronomical images. One challenge presented by colourized images is the lack of edges and shapes within astronomical objects, which is difficult for the human mind to understand. Sweitzer [37] found that when an audience is first shown a simple line drawing of the object before seeing the actual image, they are more likely to understand the concept. Additionally, it is often difficult for people to grasp the immense scales and understand the locations of objects in a sky image. Visualization tools such as SkyView [24], Google Sky [9], and WorldWide Telescope [34] can help users of any background navigate the observable universe using real astronomy data taken directly from telescope archives. Li et al. [21] provide a summary of these tools. The WorldWide Telescope Ambassadors project provides examples of "sky tours"[3] for such pedagogical purposes as explaining the difference between astronomy and astrology, learning to locate the Big Dipper, and many others.

4.3 Astronomy Data and Data-Sharing Example

The paper "A Catalog of Mid-Infrared Sources in the Extended Groth Strip" [2] provides a concrete example of data sharing in astronomy. In the terminology of the field, this is a "data paper" that describes a specific set of observations, subsequent processing, and characteristics of the resulting data set for reference by professional astronomers.[4] Here the observations are of a "blank field"—an area in the sky that has no bright celestial sources and can therefore be used to study faint, distant galaxies—with the Spitzer Space Telescope's Infrared Array Camera (IRAC) instrument [12]. The raw data comprised 18,924 individual digital images (each with 65,536 pixels and 16-bit grayscale depth) made in four wavelengths of infrared light. The IRAC camera viewed only a small portion of sky at once, so the telescope was offset between images to map a larger area. After correction for electronic and optical artifacts, the raw data were combined into four larger (about 90 MByte) 'mosaic' images, one per wavelength band. An example is shown in Fig. 3. The paper provides a link from which these images can be downloaded; the raw data are also available via the mission's online archive.

Processing the raw data is only the first step. The paper describes the properties of the mosaic images, including scatter in the values of pixels that do not contain celestial objects ('sky noise') and appearance of stars ('point spread function'); understanding these is important for measuring properties of objects that appear in the image. Specialized software was used to analyze the image pixels, identifying about 57,000 individual stars and galaxies and measuring each object's brightness, shape, and location. Extensive analysis of the precision and accuracy of these measurements was carried out to quantify their uncertainties. The resulting catalog of measurements is described in detail in the paper text. The complete catalog data are included as part of the paper's online supplementary data and available via the NED and VizieR databases. The paper concludes with a brief exploratory data analysis describing the

---

[3] https://wwtambassadors.org/tours.

[4] An example of sharing a related dataset with the public is a press release describing Hubble Space Telescope observations of the same region of sky, containing interactive images and explanatory text [36].

Fig. 3 Extended Groth Strip sky area as seen by the IRAC camera on the Spitzer Space Telescope. This is a negative image, where darker shading indicates more light received. The long image is the full mosaic image in mid-infrared (3.6 micron) light while insets show cutouts in each of the four observed wavelength bands. *Credit* Barmby et al. [2], © American Astronomical Society. Reproduced with permission

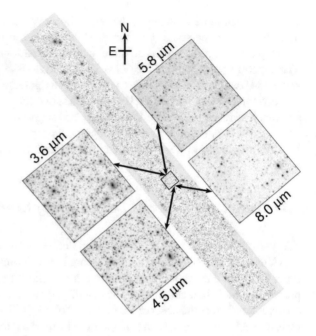

distributions of catalog quantities. Since publication, this work has been cited 82 times, in many cases by researchers unaffiliated with the original team who used the images or catalogs for their own analyses.

5　Data Challenges in Astronomy

The dataset in the above example is quite modest in size compared to the data volumes produced by current instruments. For example, the third data release from the *Gaia* mission contains measurements of 1.8 billion objects covering the entire sky. Recent improvements in telescope and detector technologies will permit astronomers to image more of the universe in better quality than ever before. Sky surveys that image all the objects in a region of the sky visible in a given waveband have increased in volume from gigabytes—often distributed on CD-ROMs in the 1990s—to petabytes today. While advances in data collection are expected to advance scientific understanding, the immense volumes of data will be a challenge for processing and analysis. To address this challenge, the field of astroinformatics has evolved as a discipline, providing formal data science methods for dealing with astronomy datasets [6]. Contributions from citizen scientists are also valuable for visually interpreting data. Both data scientists and citizen scientists often have not been formally trained in astronomy, so it is important for astronomers to effectively communicate the nature of the data that is being investigated.

Beyond sheer volume, data fusion represents an additional challenge in astronomical data analysis. Combining large datasets from multiple sources, such as sky surveys made with different telescopes, is difficult because astronomical data are inherently 'unlabelled': the visible light and radio waves arriving at Earth from a particular sky position are not guaranteed to originate from the same physical object. While prior knowledge and additional measurements can assist astronomers in combining and interpreting data from multiple sources, there is usually no way to get a "ground truth". Methods based too heavily on prior knowledge run the risk of discarding the unusual and rare objects that sky surveys are intended to detect [4].

6 Conclusions

Astronomical observations provide the raw data that enable discoveries. Astronomical imaging data, in particular, are a powerful tool to illustrate and explain scientific discoveries and facilitate public participation in and appreciation for science. The aesthetic qualities of astronomical images make astronomical data visually appealing to non-astronomers and attract "citizen scientists" to the field. Beyond their exotic subject, astronomical data are unusual in being readily available without commercial, privacy or ethical restrictions and distributed in standardized formats that can be analyzed with community-developed open-source software. Astronomical data pose unique challenges and offer unique opportunities to tell scientific stories.

Acknowledgements The authors acknowledge funding support from an NSERC Discovery Grant and Undergraduate Summer Research Award.

References

1. Arcand, K.K., Watzke, M., Rector, T., Levay, Z.G., DePasquale, J., Smarr, O.: Processing color in astronomical imagery. Stud. Media Commun. **1**(2) (2013). https://doi.org/10.11114/smc.v1i2.198
2. Barmby, P., Huang, J.-S., Ashby, M.L.N., Eisenhardt, P.R.M., Fazio, G.G., Willner, S.P., Wright, E.L.: A catalog of mid-infrared sources in the extended Groth strip. Astrophys. J. Suppl. Ser. **177**(2), 431–445 (2008). https://doi.org/10.1086/588583
3. Barmby, P.: Astronomical observations: a guide for allied researchers. Open J. Astrophys. **2**(1) (2019). https://doi.org/10.21105/astro.1812.07963
4. Baron, D.: Machine learning in astronomy: a practical overview. Preprint retrieved from http://arxiv.org/abs/1904.07248 (2019)
5. Berriman, G.B., IVOA Executive Committee, IVOA Technical Coordination Group, IVOA Community: The international virtual observatory alliance (IVOA) in 2020. Preprint retrieved from http://arxiv.org/abs/2012.05988 (2020)
6. Borne, K.D.: Astroinformatics: data-oriented astronomy research and education. Earth Sci. Inf. **3**(1–2), 5–17 (2010). https://doi.org/10.1007/s12145-010-0055-2

7. Chrysostomou, A., Taljaard, C., Bolton, R., Ball, L., Breen, S., & van Zyl, A. (2020). Operating the square kilometre array: the world's most data intensive telescope. In: Observatory Operations: Strategies, Processes, and Systems VIII, vol. 11449, p. 114490X. https://doi.org/10.1117/12.2562120

8. Collins, K.A., Kielkopf, J.F., Stassun, K.G., Hessman, F.V.: AstroImageJ: image processing and photometric extraction for ultra-precise astronomical light curves. Astron. J. **153**(2), 77 (2017). https://doi.org/10.3847/1538-3881/153/2/77

9. Connolly, A., Scranton, R., Ornduff, T.: Google Sky: a digital view of the night sky. In: Gibbs, M.G., Barnes, J., Manning, J.G., Partridge, B. (eds.) Preparing for the 2009 International Year of Astronomy: A Hands-On Symposium, vol. 400, p. 96 (2008)

10. Djorgovski, S.G., Williams, R.: Virtual observatory: from concept to implementation. In: Kassim, N.E., Perez, M.R., Junor, W., Henning, P.A. (eds.) From Clark Lake to the Long Wavelength Array: Bill Erickson's Radio Science: Astronomical Society of the Pacific Conference Series, vol. 345, p. 517 (2005)

11. English, J.: Canvas and cosmos: Visual art techniques applied to astronomy data. Int. J. Mod. Phys. D **26**(4), 1730010 (2017). https://doi.org/10.1142/S0218271817300105

12. Fazio, G.G., et al.: The Infrared Array Camera (IRAC) for the Spitzer space telescope. Astrophys. J. Suppl. Ser. **154**(1), 10–17 (2004). https://doi.org/10.1086/422843

13. Collaboration, G.: The Gaia mission. Astron. Astrophys. **595**, A1 (2016). https://doi.org/10.1051/0004-6361/201629272

14. Gaudet, S.: CADC and CANFAR: extending the role of the data centre. In: Science Operations 2015: Science Data Management, vol. 1 (2015). https://doi.org/10.5281/zenodo.34641

15. Gilda, S., Lower, S., Narayanan, D.: MIRKWOOD: fast and accurate SED modeling using machine learning. Astrophys. J. **916**(1), 43 (2021). https://doi.org/10.3847/1538-4357/ac0058

16. Gwyn, S., Willott, C., Kavelaars, J., Durand, D., Fabbro, S., Bohlender, D., Gaudet, S., Dowler, P., Jenkins, D.: Multi-archive query at the CADC: one-stop shopping for the world's astronomical data. In: Ballester, P., Ibsen, J., Solar, M., Shortridge, K. (eds.) *Astronomical Data Analysis Software and Systems XXVII*, vol. 522, p. 85 (2020)

17. Heitmann, K., Finkel, H., Pope, A., Morozov, V., Frontiere, N., Habib, S., Rangel, E., Uram, T., Korytov, D., Child, H., Flender, S., Insley, J., Rizzi, S.: The outer rim simulation: a path to many-core supercomputers. Astrophys. J. Suppl. Ser. **245**(1), 16 (2019). https://doi.org/10.3847/1538-4365/ab4da1

18. Hudec, R.: Astronomical photographic data archives: recent status. Astron. Nachr. **340**(7), 690–697 (2019). https://doi.org/10.1002/asna.201913676

19. Ivezić, Ž, et al.: LSST: from science drivers to reference design and anticipated data Products. Astrophys. J. **873**(2), 111 (2019). https://doi.org/10.3847/1538-4357/ab042c

20. Joye, W.: SAOImageDS9/SAOImageDS9 v8.0.1. (2019). https://doi.org/10.5281/zenodo.2530958

21. Li, S. et al.: The vigorous development of data driven astronomy education and public outreach (DAEPO). In: Ros, R.M., García, B., Gullberg, S.R., Moldón J., Rojo, P. (eds.) *Education and Heritage in the Era of Big Data in Astronomy*, vol. 367, pp. 199–209 (2021). https://doi.org/10.1017/S1743921321000594

22. Marshall, P.J., Lintott, C.J., Fletcher, L.N.: Ideas for citizen science in astronomy. Ann. Rev. Astron. Astrophys. **53**, 247–278 (2015). https://doi.org/10.1146/annurev-astro-081913-035959

23. Mazzarella, J.M., NED Team: Evolution of the NASA/IPAC Extragalactic Database (NED) into a data mining discovery engine. In: Brescia, M., Djorgovski, S.G., Feigelson, E.D., Longo, G., Cavuoti, S. (eds.) *Astroinformatics*, vol. 325, pp. 379–384 (2017). https://doi.org/10.1017/S1743921316013132

24. McGlynn, T., Scollick, K.: SkyView. In: Crabtree, D.R., Hanisch, R.J., Barnes, J. (eds) *Astronomical Data Analysis Software and Systems III*, vol. 61, p. 34 (1994)

25. Neronov, A.: Introduction to multi-messenger astronomy. J. Phys. Conf. Ser. **1263**, 012001 (2019). https://doi.org/10.1088/1742-6596/1263/1/012001

26. Nikutta, R., Fitzpatrick, M., Scott, A., Weaver, B.A.: Data Lab-A community science platform. Astron. Comput. **33**, 100411 (2020). https://doi.org/10.1016/j.ascom.2020.100411

27. Novacescu, J., Peek, J.E.G., Weissman, S., Fleming, S.W., Levay, K., Fraser, E.: A model for data citation in astronomical research using Digital Object Identifiers (DOIs). Astrophys. J. Suppl. Ser. **236**(1), 20 (2018). https://doi.org/10.3847/1538-4365/aab76a

28. Ochsenbein, F., Bauer, P., Marcout, J.: The VizieR database of astronomical catalogues. Astron. Astrophys. Suppl. Ser. **143**, 23–32 (2000). https://doi.org/10.1051/aas:2000169

29. Pence, W.D., Chiappetti, L., Page, C.G., Shaw, R.A., Stobie, E.: Definition of the flexible image transport system (FITS), version 3.0. Astron. Astrophys. **524**, A42 (2010). https://doi.org/10.1051/0004-6361/201015362

30. Pepe, A., Goodman, A., Muench, A., Crosas, M., Erdmann, C.: How do astronomers share data? Reliability and persistence of datasets linked in AAS publications and a qualitative study of data practices among US astronomers. PLoS One **9**, 104798 (2014). https://doi.org/10.1371/journal.pone.0104798

31. Pössel, M.: A beginner's guide to working with astronomical data. Open J. Astrophys. **3**(1), 2 (2020). https://doi.org/10.21105/astro.1905.13189

32. Rector, T.A., Levay, Z.G., Frattare, L.M., Arcand, K.K., Watzke, M.: The aesthetics of astrophysics: how to make appealing color-composite images that convey the science. Publ. Astron. Soc. Pac. **129**(975), 058007 (2017). https://doi.org/10.1088/1538-3873/aa5457

33. Reis, I., Baron, D., Shahaf, S.: Probabilistic random forest: a machine learning algorithm for noisy data sets. Astron. J. **157**(1), 16 (2019). https://doi.org/10.3847/1538-3881/aaf101

34. Rosenfield, P., Fay, J., Gilchrist, R.K., Cui, C., Weigel, A.D., Robitaille, T., Otor, O.J., Goodman, A.: AAS WorldWide telescope: a seamless, cross-platform data visualization engine for astronomy research, education, and democratizing data. Astrophys. J. Suppl. Ser. **236**(1), 22 (2018). https://doi.org/10.3847/1538-4365/aab776

35. Scroggins, M., Boscoe, B.M.: Once FITS, always FITS? Astronomical infrastructure in transition. IEEE Ann. Hist. Comput. **42**(2), 42–54 (2020). https://doi.org/10.1109/MAHC.2020.2986745

36. Space Telescope Science Institute: Hubble pans across heavens to harvest 50,000 evolving galaxies [Press release]. https://hubblesite.org/contents/news-releases/2007/news-2007-06.html (2007)

37. Sweitzer, J.S.: Strategies for presenting astronomy to the public. Int. Astron. Union Colloq. **105**, 336–339 (1990). https://doi.org/10.1017/S025292110008708X

38. The Astropy Collaboration: The Astropy project: building an open-science project and status of the v2.0 core package. Astron. J. **156**(3), 123 (2018). https://doi.org/10.3847/1538-3881/aabc4f

39. Watson, A.A.: The discovery of Cherenkov radiation and its use in the detection of extensive Air Showers. Nucl. Phys. B Proc.Suppl. **212–213**, 13–19 (2011). https://doi.org/10.1016/j.nuclphysbps.2011.03.003

40. Weekes, T.C., et al.: Veritas: the very energetic radiation imaging telescope array system. Astropart. Phys. **17**(2), 221–243 (2002). https://doi.org/10.1016/s0927-6505(01)00152-9

41. Weijmans, A.-M., Blanton, M., Bolton, A.S., Brownstein, J., Raddick, M.J., Thakar, A.: The challenges of a public data release: behind the scenes of SDSS DR13. In: Molinaro, M., Shortridge, K., Pasian, F. (eds.) *Astronomical Data Analysis Software and Systems XXVI,* vol. 521, p. 177. (2019)

42. Wenger, M., Ochsenbein, F., Egret, D., Dubois, P., Bonnarel, F., Borde, S., Genova, F., Jasniewicz, G., Laloë, S., Lesteven, S., Monier, R.: The SIMBAD astronomical database. The CDS reference database for astronomical objects. Astron. Astrophys. Suppl. Ser. **143**, 9–22 (2000). https://doi.org/10.1051/aas:2000332

43. Zwinkels, J.: Light, electromagnetic spectrum. In: Luo, R. (ed.) *Encyclopedia of Color Science and Technology*, pp. 1–8. Springer, Berlin, Heidelberg (2015). https://doi.org/10.1007/978-3-642-27851-8_204-1

P. Barmby is a professor in the Department of Physics and Astronomy at Western University, with research interests in multiwavelength observations of galaxies and astroinformatics.

S. Wong is an M.Sc. student in the Department of Physics at McGill University, with research interests in very high energy gamma-ray astrophysics.

Beyond Translation: An Overview of Best Practices for Evidence-Informed Decision Making for Public Health Practice

D. L. Schanzer, J. Arino, A. Asgary, N. L. Bragazzi, J. M. Heffernan,
B. T. Seet, E. W. Thommes, J. Wu, and Y. Xiao

Abstract The literature on best practices for evidence-informed decision-making has seen considerable growth from both knowledge users tasked with assessing the quality of the evidence and knowledge creators wishing to make a stronger contribution to evidence-based decisions. The knowledge translation process is highly dependent on the quality of the original research study, the completeness of the reporting, and the cross-discipline accessibility. The aim of this chapter is to introduce scientists

D. L. Schanzer (✉)
Ottawa, ON, Canada
e-mail: Dena.Schanzer@gmail.com

J. Arino
Department of Mathematics and the Data Science Nexus, University of Manitoba, Winnipeg, MB, Canada
e-mail: Julien.Arino@umanitoba.ca

A. Asgary
Disaster and Emergency Management, York University, Toronto, ON, Canada
e-mail: asgary@yorku.ca

N. L. Bragazzi · J. Wu
Department of Mathematics and Statistics, York University, Toronto, ON, Canada
e-mail: bragazzi@yorku.ca

J. Wu
e-mail: wujh@yorku.ca

J. M. Heffernan
Centre for Disease Modelling, Department of Mathematics and Statistics, York University, Toronto, ON, Canada
e-mail: jmheffer@yorku.ca

B. T. Seet
Medical Affairs, Sanofi (Vaccines), Toronto, ON, Canada
e-mail: Bruce.Seet@utoronto.ca

Department of Molecular Genetics, University of Toronto, Toronto, ON, Canada

E. W. Thommes
Modeling, Epidemiology and Data Science Group of Sanofi Canada, Toronto, ON, Canada
e-mail: Edward.Thommes@sanofi.com

Department of Mathematics and Statistics, University of Guelph, Guelph, ON, Canada

© The Author(s), under exclusive license to Springer Nature Switzerland AG 2023
D. G. Woolford et al. (eds.), *Applied Data Science*, Studies in Big Data 125,
https://doi.org/10.1007/978-3-031-29937-7_3

27

interested in using their statistical, analytical, mathematical, or modelling skills to contribute evidence for evidence-informed decisions in public health to the various guideline systems used in the knowledge translation process. As these guideline systems are extensive, we have provided only an overview, highlighting recommendations of potential interest to researchers reporting statistical estimates, analytical results, or modelled output. We have also included a few references to published reporting recommendations by these analytical groups. Knowledge translation does not end with a policy decision. Public health messaging is needed to inform and often persuade the general public to take the appropriate action. We have included a discussion on public communication, as media coverage of research studies can often be traced to the abstracts of the original study.

Keywords Report writing for data translation · Best practices for evidence-informed decision-making · Evaluating the quality of evidence · Novel analytical methods · Forecasting and modelled output · Public health messaging for communication and persuasion

1 Introduction

The last decade has witnessed considerable effort aiming to improve communication and collaboration between disciplines. In the area of health research, the Canadian Institute of Health Research (CIHR) was established in 2000 with a mandate to excel in the creation of new knowledge and its translation into improved health outcomes [4]. CIHR views knowledge translation as including all steps from the creation of new knowledge, by knowledge creators, to its application by knowledge users. Disciplines associated with knowledge creation have also made efforts to promote collaboration with knowledge users. For example, the Statistical Society of Canada (SSC) established the Data Science and Analytics Section with the aim of advancing Data Science and Analytics broadly, and strengthening the role of statistical science in enabling evidence-based decisions and communicating and disseminating domain-informed results. The recent publication of "Ten simple rules for effective statistical practice" [19] has useful suggestions in line with this aim.

The aim of this chapter is to introduce scientists interested in using their statistical or mathematical skills to contribute evidence for evidence-informed decisions

J. Wu
NSERC/Sanofi-York Industrial Research Chair in Vaccine Mathematics, Modelling and Manufacturing, Toronto, ON, Canada

Y. Xiao
Department of Mathematical Sciences, University of Cincinnati, Cincinnati, OH, USA
e-mail: xiaoyu@ucmail.uc.edu

J. Arino · A. Asgary · N. L. Bragazzi · J. M. Heffernan · B. T. Seet · E. W. Thommes · J. Wu · Y. Xiao
NSERC/Sanofi-York Industrial Research Chair Disease Modeling Group, Toronto, ON, Canada

in public health to the various guideline systems contributing to the knowledge translation process. These guidelines are typically written by multi-disciplinary committees with a focus on the knowledge users' perspective. The term evidence-informed decision-making implies that the decision-makers are expected to rely on their public health expertise to integrate all relevant factors, including evidence-based research studies, into any conclusions or recommendations. As decision-makers in public health are tasked with evaluating the harms and benefits of an intervention as well as the costs, cost-effectiveness, available resources, and the community or political climate, the disciplines of statistics, mathematical-modelling, and health-economics figure prominently in creating the evidence basis. Statistical methods are used in most subject matter domains to separate signal from noise, provide reliable estimates, to quantify the uncertainty of these statistical estimates, and to make inferences. Of note, publication criteria for most medical or health-science journals require that the statistical analysis be performed appropriately and rigorously. Often direct evidence of the potential effect of an intervention is not available or the actual disease burden is unknown. Mathematical models are used to bridge this gap using estimates taken from many studies and mathematical formulas, along with assumptions where data is not available. In this way, the models can account for disease progression, or transmission in infectious diseases, in describing the disease burden. Experts in health-economics incorporate cost considerations and provide decision-makers with estimates of the cost-effectiveness of an intervention.

Once a policy decision is reached, public health officials look for persuasive explanations that support the recommended policy. These more intuitive explanations are used for consensus building among health-care officials who are tasked with implementing the new policy. They will also be used to target the general public if behavioural changes are needed on their part, thus adding behavioural-science and risk-communication to the list of helpful interdisciplinary skills. Daniel Kahneman's book "Thinking, Fast and Slow" [18] offers insight into why humans struggle to think statistically and prefer to think intuitively, providing a rich source of insight on irrational decision-making. He hypothesizes that there are two modes of thought: "System 1" which is fast, instinctive, and emotional; "System 2" which is slower, more deliberative, and more logical. We often substitute an easy question for a more difficult one, so that System 1 can provide a fast answer based on a heuristic. Unless we become skeptical, System 2 will lazily endorse the conclusion without bothering to supervise. Little progress would be made if System 2 questioned everything. However, once we become skeptical and engage System 2, we can become focused on a problem and tune out other stimuli.

Likewise, throughout this chapter, we have tried to appeal to the intuitive reasoning of our readers—hoping to pique their interest in following up with the more technical details.

2 Scientific Writing and Author Guidelines for Scientific Journals

While the evidence-informed decision-making process is usually initiated with the identification of a research question and a literature review, this ignores the knowledge creation process and the importance of having high-quality studies that report findings in a way that facilitates a critical evaluation. Author guidelines and peer-review contribute to this process.

The structure of scientific articles has evolved over many years into a standardized structure known as Introduction/Methods/Results and Discussion (IMRD) [36]. This structured style facilitates finding relevant information as needed on the part of the users of the evidence, and serves as a general template for reporting study results.

As most author guidelines require that the conclusions be supported by the results presented, it is important to identify a specific research question in the introduction and link the research goal, methods, and results to the conclusions. When describing the methods, author guidelines usually require that methods be described in enough detail that another researcher with access to the data could reproduce the results.

The description of the analytical method is usually limited to a couple of paragraphs, with the methodological details provided as a reference or in an appendix. The methods section also includes a description of the study data and how it was created. Last's *Dictionary of Epidemiology (3rd ed)* [21], describes validity as the degree to which the inference drawn from a study is warranted. There are three primary threats to validity: bias, confounding, and chance. Of note, the dictionary lists over 30 types of bias, most referring to the underlying processes that generated the data. A review article on how to assess epidemiological studies [38] provides additional information on terminology and a discussion on assessing both internal and external validity for different study designs.

When introducing novel methods, author guidelines request additional information on utility (when should this novel method be used) and that a sub-section on empirical validation be included [31]. A reference should be provided for any theoretical or simulated validation exercise. The data-based validation exercise should include a comparison of the results for the novel method against commonly used methods or the gold standard for the specific application.

Guidelines for the reporting of study results depends on the study design and study objectives. This topic is discussed in more detail in the following section. Generally, when estimated parameters do not correspond to observable data, parameters should be converted to an observable quantity, including units of measure. Print journals often limit results to either a figure or table. Figures provide a visual interpretation which can facilitate communication, while studies that report point estimates along with a measure of precision such as a 95% confidence interval (CI) are more likely to be included in the knowledge translation documents prepared for the decision-making committee members, or used as parameter values to inform mathematical modelling studies. Providing the table as a supplementary file in an online journal may be a solution.

P-values are usually not used in the summary of evidence in a critical review [25]. The numerous misinterpretations of p-values and confidence intervals [12] have prompted the American Statistical Association (ASA) to publish a statement on p-values [37] where they recommend caution when reporting p-values, emphasize the need for careful interpretation of p-values in the context of the whole study design, and encourage the use of alternative approaches, such as those that emphasize estimation over testing. Many journals have responded with updated author guidelines that request that authors avoid solely reporting the results of statistical hypothesis testing, such as p-values [16].

The discussion section is tightly structured containing paragraphs to: summarize the main results; identify study limitations; compare results to those from other studies and state conclusions. The paragraph on study limitations should include a discussion of the potential risk of bias and potential confounders not included in the study design.

It should be noted that "conclusions about the validity of a study require wisdom and rigor to apply expert judgment based on knowledge of the subject matter and of the methodology" [32]. With new analytical methods emerging quickly, the task of assessing validity will increasingly fall to experts with skills in both the analytical and subject matter domains.

The issue of conveying study quality for mathematical modelling and health-economic studies is more complex and recommendations less developed. Typically, scenarios and sensitivity analysis are used to convey uncertainty. Reporting guidelines specific to these study designs are discussed in the next section.

3 Reporting Guidelines

An international initiative was set up in 2006 to promote good reporting practices, including the wider implementation of reporting guidelines. The EQUATOR (Enhancing the QUAlity and Transparency Of health Research) Network was motived by concerns "that deficiencies in reporting make it difficult, if not impossible, to assess how the research was conducted, to evaluate the reliability of the presented findings, or to place them in the context of existing research evidence" [35]. Many journals now request that the appropriate reporting guideline be used by authors and reviewers [16].

The EQUATOR network [9] consists of an online library of reporting guidelines and check-lists organized by study type, as well as guidelines for the reporting of statistical results. The SAMPL (Statistical Analysis and Methods in Published Literature) Guidelines [20] suggest two guiding principles: (1) describe statistical methods in enough detail to enable a knowledgeable reader with access to the original data to reproduce the published results; (2) provide enough detail that the study results can be incorporated into other analyses such as meta-analyses. Some newer additions to the network include RECORD and CHEERS (Consolidated Health Economic Evaluation Reporting Standards). The RECORD guidelines are an extension of the

STROBE guidelines for observational studies that use routinely-collected health data as the main data source. These databases provide a wealth of information that previously was very costly to obtain.

Studies based on mathematical models, including health-economic studies, pose a particular challenge for reviewers, as an empirical measure of precision, or an empirical study with which to compare results, may not available to assess the study quality. Early attempts to development guidelines on best modelling practices by the ISPOR-SMDM Modeling Task Force [5] tended to be rather technical and required an in-depth familiarity with mathematical modelling. Results of theoretical validation efforts or simulations remain mostly out of the reach of policy decision-makers.

Often, mathematical models are used to help assess harms and benefits of a potential intervention for which we have some data, but for which there are also data gaps. Models are often the only option short of a pilot study, as they can link existing data along with other assumptions derived in part from expert opinions, and provide much needed insight on potential outcomes. The CHEERS guidelines are a welcome addition, as they are written for reviewers who may have a health-science background. Though written for cost-effectiveness studies, CHEERS is also recommended for studies that report on modelled output [16].

It is worth noting that the development of the CHEERS guideline was motivated as well by commonly observed deficiencies in reporting. A central requirement in CHEERS is to provide a list of all parameters and assumptions used to inform the model. It is important that the reference is to the original study (or meta-analysis) rather than another modelling study, as many details in the original manuscript are needed to assess whether a referenced estimate is applicable for the use it is put to in the model. The range of uncertainty (95% CI) associated with each parameter is required for the sensitivity analysis. As the omission of structural assumptions can limit the quality of an assessment, suggestions to address this issue are discussed in the next section.

Cost-effectiveness studies usually include a summary measure of the net costs and benefits of an intervention over the patient's life-time, with the net benefits reported as a Quality-Adjusted Life Year (QALY) estimate. The QALY accounts for both life years gained and the improvement in quality-of-life for the rest of the patient's life. In addition to the full itemization of all assumptions, these studies should list all the harms and benefits going into the QALY estimate, as well a providing the standard disability weight for each harm and benefit. Itemizing the QALY contributions and resource requirements on an annual basis would be helpful for assessing operational considerations such as budgets, resource constraints, and the expected timing of benefits. Itemization is required so that reviewers can reproduce the QALY estimate and confirm whether all potential harms and benefits have been included. Suspicions that important harms or benefits were omitted can result in unnecessary discord among the decision-making committee.

The ISPOR-SMDM Modeling Task Force's guidelines on best modelling practices identified the issue of not accounting for operational considerations, such as short-term resource constraints, or the use of model parameters rather than actual data to describe the intervention scenarios, as one of the limitations of many modelling studies [5]. Suggested solutions include cross-disciplinary collaboration to identify possible resource constraints and the development of more complex models that include realistic time-tables for the implementation of an intervention and link model parameters to actual data.

4 Guidelines for Development of Health Policy Recommendations: Grading the Evidence

Systematic reviews provide a comprehensive overview of the available evidence relevant to a policy decision. The GRADE (Grading of Recommendations, Assessment, Development and Evaluations) approach was first published as a six-part series in 2008 [14], and has continued to expand, becoming the most widely adopted tool for grading the quality of evidence [34]. The complete set of guidelines are available online [17]. These guidelines outline the importance of framing the question, selecting appropriate outcomes and describe how to rate the quality of evidence based on risk of bias, publication bias, imprecision (or random error), inconsistency and indirectness. While most data-scientists are not involved in assessing the quality of the evidence, the GRADE guidelines provide an understanding of how a study may be assessed.

To summarize the evidence from each individual study, the GRADE approach reports the estimated impact of the intervention, usually with the lower 95% CI, and uses a 4-point scale for quality of evidence (very low, low, moderate, high). All harms and benefits associated with the intervention are itemized. After weighing all harms and benefits, operational considerations, and the quality of evidence, the recommendation is rated on its impact (strong, weak). The summary part of the GRADE system is an impressive knowledge translation tool, supported by years of experience of diverse experts, including behaviour science [13]. Additional evidence-based tools for these tasks are available online from The National Collaborating Centre for Methods and Tools (NCCMT), McMaster University, Canada [25] and the BMJ Publishing Group [1], among others. The resulting health-policy guidelines are widely disseminated in various online libraries, for example, the Public Health Agency of Canada [30], the Canadian Medical Association Infobase [7], university libraries or disease specific associations.

As the evidence base is much less developed for public health interventions compared with clinical health where randomized control trials are the gold standard, the demand for more evidence on the costs and benefits has increased, coupled with a demand for guidelines for assessing these studies. For example, the National Advisory Committee on Immunization (NACI) is updating its public health recommendation process for vaccines to include economic analyses [24], and provides online access to additional reporting and assessment guidelines.

The GRADE Working Group has recently published their 30th guideline, an overview of the GRADE approach for assessing the certainty of modelled evidence

[3]. To accommodate GRADE prinicples, the credibility of a model itself and the certainty of evidence for each of the model inputs should be assessed, for example by applying GRADE to each model input and identifying those parameters to which model outputs are most sensitive. However, the information required to assess someone else's model is often missing or difficult to obtain. To identify assumptions implicit in the model structure, the working group envisioned comparing the outputs of multiple models, or attempting to identify the 'ideal' model in order to include less obvious parameters or assumptions in the sensitivity analysis.

When the model outputs are sensitive to the model type or structure, or require inputs for which the values are unknown, a full GRADE evaluation may be premature. In this case, the model output could be viewed as hypothesis-generating, similar to how ecological studies are viewed. The identification of which parameters and model structures are responsible for the most uncertainty, through theoretical validation, simulation exercises and sensitivity analyses, would help document the most important data gaps where better-quality data is most needed. In some cases, such as for pandemics, or environmental events such as a hurricane, the data inputs can change quickly. Input parameters that are not likely to be consistent over time should be flagged when assessing inconsistency. A more complex model may be required to link the usual model inputs to available surveillance data. Once linked, thresholds could be set to alert officials when the modelled output is likely no longer reliable enough—prompting a quick reassessment. When model type seems to be responsible for substantial variation in model outputs, a direct comparison of model outputs for different model types is needed.

5   Post-Decision Consensus Building and Public Health Messaging

As the COVID-19 pandemic has illustrated, convincing the public to co-operate with a policy decision can be just as important as getting the policy recommendation right. In this phase, public health officials often look for persuasive explanations that support the recommended policy and are easily understood. While the importance of consistent messaging from politicians and scientific experts cannot be underestimated, decades of research in risk-communication shows that many factors are involved in gaining the public's compliance and that too often these messages do not work as intended. For example, before modifying their own behaviour, people need a good perception of their own risk, they need to trust the message, and they need to be empowered to take preventive measures [10].

The risk-communication research literature is large and diverse, however, as with most academic bodies of literature, it is typically out of the reach of researchers from other disciplines as well as public health officials. To bridge the gap, the

US Food and Drug Administration (FDA) published a guide to facilitate evidence-based risk-communication [10]. Risk-communication is distinguished from public-relations communication by its commitment to accuracy and its avoidance of spin. While "spin" is often associated with media consultants who develop deceptive or misleading messages to influence the public, spin in media coverage of research can often be traced to the abstracts [33]. Statements in research articles that intentionally or unintentionally overstate the beneficial effects of an intervention were found to be mainly related to misleading reporting or misleading interpretation of the study results [15]. These same issues were cited by the GRADE working group as motivating guideline development.

Solutions may lie in closer compliance with reporting guidelines among editors, reviewers, and authors. As many of us observed during the COVID-19 pandemic, health communication appears to have been designed to persuade people more than to inform them. Generally, the more certainty there is about the balance between the advantages and disadvantages of the change in behaviour, and the greater the potential for harm to others (e.g., transmission of infectious diseases or drunk driving), the more likely it is that persuasion is justified [29]. Too much spin, for example, by not disclosing uncertainties, distorts what is known, inhibits research to reduce important uncertainties, and can undermine public trust in health authorities. However, sometimes persuasion is not effective enough and mandates are used [10], for example with helmets for motorcycle or bicycle use, or COVID-19 vaccine passports to help persuade more people to get vaccinated.

A formal evaluation of the message using focus groups, as well as consulting experts in risk-communications can reduce the risks that a message backfires or undermines public trust [28]. Even early in the COVID-19 pandemic, participants of a focus group identified problems with the current public health messaging, such as inconsistency, lack of transparency, and lack of the supporting scientific data presented by a trustworthy source [11]. Of note, the participants perceived that public health officials were over confident in presenting model projections when a hopeful prediction turned out to be wrong, or lost trust in officials when predictions looked too dire or the intervention too severe. Admitting mistakes is rare, though insightful. For example, the Modelers' Hippocratic Oath, written in response to the role of Quants and their mathematical models in the 2008 stock market crash [8] reflects on behavioural biases that led the group to inadvertently put a bit of spin on their work. One of the cognitive biases that affects our decision making is the IKEA effect, where people place a disproportionately high value on products they partially created [27]. It is named after IKEA, where consumers assemble the modular furniture themselves. Confirmation bias, where one tends to search for and interpret information in a way that confirms one's prior beliefs, is one of the many cognitive biases—that is, errors in logic that arise from using personal beliefs or experiences to make quick decisions, or in Daniel Kahneman's terminology, System 1 hijacks our critical thinking process.

6 Summary

Increasingly peer-reviewers are encouraged to specifically explain whether and how the manuscript could be improved to follow the appropriate reporting guidelines more closely. As guidelines are open to interpretation, and even statistical methods are based on some assumptions, a classroom discussion is a good setting to become familiar with the relevant guidelines and to hear a range of interpretations. While, as research scientists, we may not be formally asked to do a criterial review of the evidence, some familiarity with this process provides an understanding of how the quality of the evidence as reported in our studies could be assessed and how it contributes to a policy decision. The knowledge translation process is highly dependent on the quality of the studies and the completeness of the reporting. Often access to the data and strong statistical skills are required to assess imprecision and risk of bias inherent in the methodology. If these topics are not addressed in the study report, they usually cannot be assessed in the critical review, thus risking a downgrade in the quality of the evidence. Researchers with the appropriate statistical or mathematical modelling skills and domain-specific skills can improve the quality of these important critical reviews.

New analytical methods are quickly emerging, and these pose a challenge for the critical review process. If the results from a novel method are not compared to results for the commonly used methods or the gold standard, these details are not available for translation and critical reviewers would have difficulty interpreting the study results. For modelled output, reviewers are looking at questions such as: how reasonable the assumptions are, whether the timelines for implementation of an intervention are realistic, whether input parameters are measurable, or which factors were accounted for in the model—questions identified by the ISPOR-SMDM Modeling Task as operational concerns.

Part of gaining more familiarity with reporting guidelines includes participating in group discussions, or for a more hands-on approach, peer-reviewing manuscripts, conducting a method comparison study, or a systematic review of studies that use new or different methods or modelled output. The demand for forecasts of hospital resource requirements during COVID-19 epidemic waves provides one example of the importance of familiarity with critical review and reporting criteria. A systematic review of COVID-19 forecasting studies found that half of the studies did not report the quantitative uncertainty of their predictions; 25% did not conduct an evaluation of their short-term forecast, and most did not evaluate their forecasts over a period of time that included varying epidemiological dynamics [26]. It is promising to see that there are also a number of recent studies that compared the accuracy of the forecasts of different models over an extended forecast period by using period-comparison methods such as the weighted interval score (WIS) [2], and included forecast periods over varying epidemiological dynamics [23]. Collaboration between forecasters and knowledge-users has led to insightful discussions. For example, as public health is generally interested in the peak characteristics of an epidemic wave, the need for alternative measures of accuracy is increasing being recognized. The issue is that typical error-based metrics, such the mean-squared error averaged over the full time-series can lead to poor performance in assessing the precision of predictions of the timing and magnitude of the epidemic peak [6, 22].

Spin in the media can influence the public, and can often be traceable to research abstracts. As the public generally views independent researchers more trustworthy than government officials, data-scientists, with a good understanding of the information that the reporting and grading guidelines are looking for, should be well placed to gain public trust by providing an informative rather than persuasive presentation of the supporting evidence.

Funding and Conflict of Interest Declarations BTS was employed with Sanofi (Vaccines) Canada at the time of writing; he is now employed at Novavax Inc. DLS had retired from the Public Health Agency of Canada at the time of writing. EWT is employed by Sanofi Canada. JA is partially funded by NSERC. JMH is funded by NSERC DG, NSERC-PHAC EIDM, and CIHR COVID-19. JMH has recently consulted for Sanofi and Health Canada. JMH is currently a Scientific Advisor for Canada's COVID Immunity Task Force. JW is partially funded by NSERC.

References

1. BMJ Publishing Group Limited: BMJ Best Practice (2022). https://bestpractice.bmj.com/info/toolkit/
2. Bracher, J., Ray, E.L., Gneiting, T., Reich, N.G.: Evaluating epidemic forecasts in an interval format. PLoS Comput. Biol. **17**(2), e1008618 (2021). https://doi.org/10.1371/journal.pcbi.1008618
3. Brozek, J.L., Canelo-Aybar, C., Akl, E.A., Zhang, Y., Schuneman, H.J.: for the GRADE Working Group: GRADE Guidelines 30: the GRADE approach to assessing the certainty of modeled evidence—an overview in the context of health decision-making. J. Clin. Epidemiol. **129**, 138–150 (2021). https://doi.org/10.1016/j.jclinepi.2020.09.018
4. Canadian Institutes of Health Research: Knowledge Translation. Government of Canada (2016). https://cihr-irsc.gc.ca/e/29418.html
5. Caro, J.J., Briggs, A.H., Siebert, U., Kuntz, K.M.: Modeling good research practices—overview: a report of the ISPOR-SMDM modeling good research practices task force-1. Value Health **15**(6), 796–803 (2012). https://doi.org/10.1016/j.jval.2012.06.012
6. Chakraborty, P., Lewis, B., Eubank, S., Brownstein, J.S., Marathe, M., Ramakrishnan, N.: What to know before forecasting the flu. PLoS Comput. Biol. **14**(10), e1005964 (2018). https://doi.org/10.1371/journal.pcbi.1005964
7. CMA Impact Inc.: CPG Infobase: Clinical Practice Guidelines. CMA Joule (2022). https://joulecma.ca/cpg/homepage
8. Derman, E., Wilmott, P.: The Financial Modelers' Manifesto (2009). https://www.uio.no/studier/emner/sv/oekonomi/ECON4135/h09/undervisningsmateriale/FinancialModelersManifesto.pdf
9. EQUATOR Network: The EQUATOR Network: Enhancing the QUAlity and Transparency of health Research (2022). https://www.equator-network.org/
10. Fischhoff, B., Brewer, N.T., Downs, J.S.: Communicating Risks and Benefits: An Evidence-Based User's Guide (2011). https://www.fda.gov/media/81597/download
11. Fullerton, M.M., Benham, J., Graves, A., Fazel, S., Doucette, E.J., Oxoby, R.J., Mourali, M., Boucher, J.-C., Constantinescu, C., Parsons Leigh, J., Tang, T., Marshall, D.A., Hu, J., Lang, R.: Challenges and recommendations for COVID-19 public health messaging: a Canada-wide

qualitative study using virtual focus groups. BMJ Open **12**(4), e054635 (2022). https://doi.org/10.1136/bmjopen-2021-054635

12. Greenland, S., Senn, S.J., Rothman, K.J., Carlin, J.B., Poole, C., Goodman, S.N., Altman, D.G.: Statistical tests, P values, confidence intervals, and power: a guide to misinterpretations. Eur. J. Epidemiol. **31**(4), 337–350 (2016). https://doi.org/10.1007/s10654-016-0149-3

13. Guyatt, G.H., Oxman, A.D., Schünemann, H.J., Tugwell, P., Knottnerus, A.: GRADE guidelines: a new series of articles in the Journal of Clinical Epidemiology. J. Clin. Epidemiol. **64**(4), 380–382 (2011). https://doi.org/10.1016/j.jclinepi.2010.09.011

14. Guyatt, G.H., Oxman, A.D., Vist, G.E., Kunz, R., Falck-Ytter, Y., Alonso-Coello, P., Schünemann, H.J.: GRADE: an emerging consensus on rating quality of evidence and strength of recommendations. BMJ (Clinical Research Ed.) **336**(7650), 924–926 (2008). https://doi.org/10.1136/bmj.39489.470347.AD

15. Haneef, R., Lazarus, C., Ravaud, P., Yavchitz, A., Boutron, I.: Interpretation of results of studies evaluating an intervention highlighted in google health news: a cross-sectional study of news. PLoS ONE **10**(10), e0140889 (2015). https://doi.org/10.1371/journal.pone.0140889

16. JAMA Network Open: Instructions for Authors. JAMA Network (2022). https://jamanetwork.com/journals/jamanetworkopen/pages/instructions-for-authors

17. Journal of Clinical Care GRADE Website: GRADE Series (2022). https://www.jclinepi.com/content/jce-GRADE-Series

18. Kahneman, D.: Thinking, fast and slow. Farrar, Straus and Giroux (2011)

19. Kass, R.E., Caffo, B.S., Davidian, M., Meng, X.-L., Yu, B., Reid, N.: Ten simple rules for effective statistical practice. PLoS Comput. Biol. **12**(6), e1004961 (2016). https://doi.org/10.1371/journal.pcbi.1004961

20. Lang, T.A., Altman, D.G.A.: Basic statistical reporting for articles published in biomedical journals: The "Statistical Analyses and Methods in the Published Literature" or The SAMPL Guidelines". In: Smart, P., Masisonneuve, H., Polderman, A. (eds.) Science Editors' Handbook. European Association of Science Editors (2013). https://www.equator-network.org/wp-content/uploads/2013/07/SAMPL-Guidelines-6-27-13.pdf

21. Last, J.M., and International Epidemiological Association: A Dictionary of Epidemiology, 3rd edn. Oxford Univeristy Press (1995)

22. Lutz, C.S., Huynh, M.P., Schroeder, M., Anyatonwu, S., Dahlgren, F.S., Danyluk, G., Fernandez, D., Greene, S.K., Kipshidze, N., Liu, L., Mgbere, O., McHugh, L.A., Myers, J.F., Siniscalchi, A., Sullivan, A.D., West, N., Johansson, M.A., Biggerstaff, M.: Applying infectious disease forecasting to public health: a path forward using influenza forecasting examples. BMC Public Health **19**(1), 1659 (2019). https://doi.org/10.1186/s12889-019-7966-8

23. Meakin, S., Abbott, S., Bosse, N., Munday, J., Gruson, H., Hellewell, J., Sherratt, K., Chapman, L.A.C., Prem, K., Klepac, P., Jombart, T., Knight, G.M., Jafari, Y., Flasche, S., Waites, W., Jit, M., Eggo, R.M., Villabona-Arenas, C.J., Russell, T.W., and Group, C. C.-19 W: Comparative assessment of methods for short-term forecasts of COVID-19 hospital admissions in England at the local level. BMC Med. **20**(1), 86 (2022). https://doi.org/10.1186/s12916-022-02271-x

24. National Advisory Committee on Immunization: Process for Incorporating Economic Evidence into Federal Vaccine Recommendations Stakeholder Consultation (2021). https://www.canada.ca/content/dam/phac-aspc/documents/programs/process-incorporating-economic-evidence-federal-vaccine-recommendations-stakeholder-consultation/document/economic-process-consultation.pdf

25. National Collaborating Centre for Methods and Tools: Evidence-Informed Decision Making in Public Health (2022). https://www.nccmt.ca/tools/eiph

26. Nixon, K., Jindal, S., Parker, F., Reich, N.G., Ghobadi, K., Lee, E.C., Truelove, S., Gardner, L.: An evaluation of prospective COVID-19 modeling: from data to science translation. MedRxiv, 2022.04.18.22273992 (2022). https://doi.org/10.1101/2022.04.18.22273992

27. Norton, M., Harvard, B.S., Mochon, D. (Yale), Ariely, D. (Duke): The IKEA effect: when labor leads to love. J. Consumer Psychol. **22**(3), 453–460 (2012). https://doi.org/10.1016/j.jcps.2011.08.002

28. NPHIC: Ten Communication Tactics to Combat Pandemic Messaging Fatigue. National Public Health Information Coalition (2022). https://nphic.org/news/featured-topics/697-10comm-tactics

29. Oxman, A.D., Fretheim, A., Lewin, S., Flottorp, S., Glenton, C., Helleve, A., Vestrheim, D.F., Iversen, B.G., Rosenbaum, S.E.: Health communication in and out of public health emergencies: to persuade or to inform? Health Res. Policy Syst. **20**(1), 28 (2022). https://doi.org/10.1186/s12961-022-00828-z

30. PHAC: Disease Prevention and Control Guidelines (2022). https://www.canada.ca/en/public-health/services/infectious-diseases/nosocomial-occupational-infections.html

31. PLOS ONE: Submission Guidelines (2022). https://journals.plos.org/plosone/s/submission-guidelines#loc-methods-software-databases-and-tools

32. Porta, M., and International Epidemiological Association: A Dictionary of Epidemiology, 6th edn. Oxford University Press (2014)

33. Roehr, B.: "Spin" in media coverage of research can be traced to abstracts. BMJ: Br. Med. J. **345**, e6106 (2012). https://doi.org/10.1136/bmj.e6106

34. Siemieniuk, R., Guyatt, G.H.: What is GRADE. BMJ Best Practices (2022). https://bestpractice.bmj.com/info/toolkit/learn-ebm/what-is-grade/

35. Simera, I., Moher, D., Hirst, A., Hoey, J., Schulz, K.F., Altman, D.G.: Transparent and accurate reporting increases reliability, utility, and impact of your research: reporting guidelines and the EQUATOR Network. BMC Med. **8**, 24 (2010). https://doi.org/10.1186/1741-7015-8-24

36. Sollaci, L.B., Pereira, M.G.: The introduction, methods, results, and discussion (IMRAD) structure: a fifty-year survey. J. Med. Library Assoc. JMLA **92**(3), 364–367 (2004). https://europepmc.org/articles/PMC442179

37. Wasserstein, R.L., Lazar, N.A.: The ASA statement on p-values: context, process, and purpose. Am. Stat. **70**(2), 129–133 (2016). https://doi.org/10.1080/00031305.2016.1154108

38. Zaccai, J.H.: How to assess epidemiological studies. Postgraduate Med. J. **80**(941), 140 LP–147 (2004). https://doi.org/10.1136/pgmj.2003.012633

Concern for Self-Health During the COVID-19 Pandemic in Canada: How to Tell an Intersectional Story Using Quantitative Data?

Laila Rahman(iD)

Abstract Rooted in Black feminism, intersectionality theory entered critical legal studies and travelled to public health and beyond. This chapter demonstrates one of many ways to apply intersectionality theory using a *descriptive intercategorical* approach to quantitative data. In so doing, I attempt to tell an intersectional story to make visible the intersectional inequalities for Canadians' concerns for self-health during the first wave of COVID-19 pandemic. For pedagogical purposes, I share a subset of Statistics Canada's COVID-19 Impacts Survey 2020 dataset of 239,143 participants and Stata code to encourage students to practice estimating intersectional outcomes and ask questions to explicate health inequalities. Although interrogating the systems of power is critical, this project does not statistically analyse but draws on the literature to discuss how interacting power structures might shape intersectional peoples' experiences. In addition, the analysed dataset is not representative of the Canadian population. Nonetheless, it might be helpful to showcase a case study on introductory-level quantitative intersectionality research. I hope, despite these limitations, this case study and the pedagogical tools will contribute to greater access to intersectionality research, generating a cadre of *intersectionality data translators* in public health.

1 Introduction

After a state of plague was declared and the town gates were shut down, "every one of us realized that *all... were, so to speak, in the same boat*" [11, p. 66]. Like Camus's protagonist, when the novel coronavirus disease 2019 (COVID-19) emerged, some experts thought that all Canadians were in this pandemic together [3], which was far from the truth [2]. In this chapter, I examined the patterns of inequalities in COVID-related concern for self-health across not only single but different intersectional

L. Rahman (✉)
Department of Epidemiology and Biostatistics, Schulich School of Medicine and Dentistry, University of Western Ontario, London, ON, Canada
e-mail: lrahman8@uwo.ca

© The Author(s), under exclusive license to Springer Nature Switzerland AG 2023
D. G. Woolford et al. (eds.), *Applied Data Science*, Studies in Big Data 125,
https://doi.org/10.1007/978-3-031-29937-7_4

social locations by applying intersectionality theory. Rooted in Black feminism, intersectionality theory entered critical legal studies [15] and travelled to public health and beyond [5]. This chapter demonstrates one way to translate Statistics Canada's [39] dataset to make visible the intersectional inequalities in Canada. In so doing, I attempt to tell an intersectional story using quantitative data. I share the data solution lifecycle in forming research questions, accessing the dataset, conducting data analysis, and interpreting results. For pedagogical purposes, the analytical dataset and Stata [38] code are published online [33]. However, it is important to note that this dataset does not represent the Canadian population. Despite this limitation, I hope this case study and the pedagogical tools will contribute to greater interest and access to intersectionality research, generating a cadre of *intersectionality data translators* in public health.

I begin by defining intersectionality and describing approaches to doing quantitative intersectionality research. Then, I present a case study to show *what* quantitative intersectionality can add to the field; and discuss the *limits* of these analyses and what I did to avoid normalizing inequalities [4, 27]. I close with questions and welcome readers to formulate inquiries to advance intersectionality research.

2 Intersectionality Theory

Drawing on Black feminism and critiquing the silos of critical race and feminist theories, with the former focusing on Black men in fighting against racism and the latter focusing on white women in fighting against sexism, intersectionality theory highlights the experiences of Black women who occupy in-between marginal spaces [14, 15]. Such blind spots are created by interacting systems of power such as racism and sexism, making Black women invisible so that their problems are not recognized as distinct from that of Black men or white women, who are erroneously considered 'the' representatives of Black race and women gender [12]. Refreshingly, intersectional thinking has provided a lens to understand different interacting systems of power and the complexity of the social world and human experiences to exact change [12, 13]. As Collins and Bilge [13] identified six core ideas of intersectionality—social inequalities, social context, social justice, power, relationality, and complexity in the field of sociology, Bowleg [8] pointed to intersectionality's *three core tenets* in public health. These tenets guide researchers to:

(1) Consider social identities, locations, or positions as multi-dimensional, interdependent, multiple, and intersecting;
(2) Start analysis with historically or currently oppressed and marginalized groups;
(3) Consider how the social identities or locations at the individual- or micro-level (e.g., gender, race, socio-economic status) interact with the systems and structures at the macro-level (e.g., patriarchy/sexism/misogyny, racism, poverty), producing inequitable health outcomes.

3 Quantitative Intersectionality Approaches

McCall [29] introduced a typology that included (a) anticategorical, (b) intercategorical, and (c) intracategorical intersectional approaches. An *anticategorical* approach questions categories [29]; therefore, research using this approach shows low discriminatory accuracy of models as evidence of the absence of categories (see [18]). An *intracategorical* approach focuses on differences *within* a specific marginalized group. Examining minority stress among lesbian, gay, and bisexual people of colour is an example of inquiries using this approach (see [35]). In contrast, an *intercategorical* approach examines differences and sameness across existing, socially constructed intersectional groups [29]. A study examining chronic obstructive pulmonary disease across intersections of age, gender, income, education, civil, and migration status is an example of such research (see [19]).

Intersectionality research can also be divided into (a) quantitative, (b) qualitative, and (c) mixed methods. Bauer and Scheim further divided quantitative intersectionality research into (a) descriptive and (b) analytic approaches [6]. The former performs statistical analysis to estimate and compare the outcome patterns across intersections, while the latter presents statistical analyses of the causal processes leading to such results. Both approaches draw on the literature on interacting systems of power, but analytical research measures these systems and processes, going beyond the comparisons across intersections. A study examining intimate partner violence across different intersections is an example of descriptive analysis (see [34]), while another study investigating the mediating effect of discrimination on distress across intersections is analytical research (see [6]).

Although applications of quantitative intersectionality research have increased rapidly since 2001, most studies (such as this chapter's case study) are intercategorical and descriptive, and a majority of them use conventional approaches (e.g., multiple regression), while others use emerging techniques (e.g., multilevel analysis of individual heterogeneity and discriminatory accuracy, MAIHDA [5]).

Qualitative researchers have critiqued quantitative intersectionality research by equating quantitative analyses to positivism which in the past produced dubious knowledge, for example, analysing men-only samples to generalize findings for the total population [1, 37]. In the 1920s, positivist research emphasized mathematics-like precision, although, in the social sciences, hypotheses can never be fully proved or disproved [36]. Quantitative scholars refute these concerns because quantitative research can be critical and non-positivist by following intersectionality theory's core tenets and attending to complexity and power [8, 12, 21, 29]. Qualitative intersectionality researchers can, undoubtedly, present marginalized groups' compelling and complex narratives in their own voices [9, 10], which is impossible in quantitative analysis. Quantitative researchers, however, can tell an intersectional story because intersectional realities, although complex, are nonetheless *patterned*, making them amenable to explanation [29]. Quantitative inquiries can identify *complex outcome patterns* across intersections using large data, producing new insights to understand and address inequities [4]. Intersectionality researchers, however, need to avoid analysing the poor health outcomes without discussing the underlying power structures because this might reify categories [4].

4 What Does Quantitative Intersectionality Add to Public Health?

Intersectionality's core idea of social justice is consistent with public health's goal to address social inequalities [42]. Although the effects of social locations on health have long been acknowledged, risk factor epidemiology adjusts for them instead of making them subjects of inquiry [25]. Social epidemiology emerged as a field within public health in the early 1960s by critiquing this approach [7]. However, many studies examining social locations consider them independently, with one category (e.g., race/ethnicity, a *single-category location*) added to others (e.g., gender, class), creating multiple but independent categories (ethnicity + gender + class). This analytical approach is known as the *additive approach* [21].

Intersectionality scholars have critiqued this approach because people do not live single-category lives [22]. Although social categories might be produced by distinct systems of power (e.g., ageism, racism, sexism, classism), they interact and mutually constitute each other [12, 15]. Therefore, an *intersectional approach* allows for examining complex intersections. For example, in examining Black women's particular health outcomes, an additive approach considers (a) the experiences of 'Blacks' *plus* (b) the experiences of 'women.' In doing so, this approach, on the one hand, conflates women's and men's experiences within 'Blacks,' and on the other hand, Blacks' and whites' experiences within 'women.' To reduce health inequalities, it is, therefore, important to use an intersectional lens that does not conflate categories but meets people at intersections where they are to cater to their unique needs, for example, as 'Black women' instead of essentializing them as 'women' (Fig. 1).

Fig. 1 How categories are treated in an additive vs. an intersectional approach (see the top vs. the bottom panel)

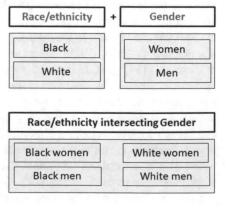

I now turn to a case study highlighting the difference between an *additive analytical* and a *descriptive, intercategorical intersectional* approach.

.

5 Case Study: Concerns During the COVID-19 Pandemic in Canada

Worldwide, COVID-19 has caused millions of deaths and morbidities and triggered pandemic-related concerns [26, 43]. As Canada was experiencing the first wave of the COVID-19 pandemic, Statistics Canada conducted an impacts survey online [39]. I have used their data to examine Canadians' concerns for self-health across intersections. To manage complexity for pedagogical purposes, I restricted my focus to participants' four social categories: age, gender, income, and community type. Considering the social structures and systems of power, including ageism, sexism, the neoliberal economic system, and rural–urban divide, I assumed oppressed social locations to be older age, women, nonsecure income, and living in rural or small urban communities, these formed the starting point for my analyses in the Canadian context.

5.1 Research Objective and Questions

This project's objective was to advance intersectionality scholarship by drawing a contrast between additive and intersectional approaches to examine the COVID-19 impact on intersectional groups of Canadians. Using an additive approach, I asked: How do Canadians' concerns for self-health vary across age, gender, income, and community locations? Next, using an intersectional lens, I asked: *To what extent do concerns for self-health vary across different social locations at intersections of age, gender, income, and community locations?* Comparing responses to these questions allowed me to show the difference between additive and intersectional approaches.

5.2 Methods

Data source This case study uses Statistics Canada's crowdsourcing data, collected from April 3 to 23, 2020, to understand the impacts of the COVID-19 pandemic in Canada [39].

Study designs The first research question using an additive approach employed a quantitative analytical design because, in standard research methods parlance, comparative research is considered analytical. The second question using an intersectional approach, employed a quantitative, descriptive, intercategorical intersectional design because comparisons across intersections are the starting point in intersectionality research.

Survey This study used secondary, cross-sectional data from a non-probabilistic sample [26]. The survey was *cross-sectional* because the data about participants' social locations (explanatory variables) and their concern for self-health (outcome variable) were collected at the same time [40]. The sample was *non-probabilistic* because the respondents self-selected online [28, 40].

The sample comprised 242,519 Canadians aged 15 years and older [39]. I excluded 3376 (1.4% of the sample) missing cases in the outcome and explanatory variables to arrive at an analytical sample of 239,143.

Outcome variable Participants' concern for their health due to the COVID-19 pandemic was the outcome variable. Participants who were *extremely concerned* were coded as 1, and those who were very, somewhat, and not at all concerned were coded as 0.

Explanatory, social location variables Intersectional categorical variables included participants' age, gender, income, and community type. Using these four variables, each with 2 levels, generated $(4 \times 2 =)$ 8 primary categories which allowed creating $(2 \times 2 \times 2 \times 2 =)$ 16 intersectional social locations. This categorization attends to complexity by doubling the number of social locations, leading to 120 pairwise comparisons (combination, $c(n, r) = n!/r!(n—r)! = 16!/2!(16—2)!$).

Participant ages in years were dichotomized, with 40 years and older coded as 1 and 15 to 39 years, 0. Thus, the older age group represents the seniors, baby boomers, and generation X, while the younger age group represents the millennials and generation Z [16]. Gender variable was coded 1 for women, otherwise, 0 for men. Nonsecure income variable was created by coding participants who stated that they would have a major or moderate impact to meet financial obligations or essential needs as 1, otherwise, 0. Community size variable was created with two categories—communities, where less than 500,000 people lived, were considered small communities and coded 1; otherwise, they were considered large communities and coded 0.

Thus, age, gender, income, and community size variables, each with two response categories, coded as 1 and 0 (i.e., *dummy coding*), allowed comparisons between any two single-category locations within the same category, for example, young vs. old within age category and men vs. women within gender category for additive analysis. Taken together, these four variables represented eight single-category locations: (1) old, (2) young, (3) women, (4) men, (5) nonsecure income, (6) secure income, (7) small communities and (8) large communities. These binary variables were cross-classified to generate one intersectional location variable with *16 values*, coded from 0 to 15, each representing one intersection.

Confounding variable Province variable was associated with community size and the outcome but did not mediate this relationship. Therefore, I considered this variable a confounding factor.

5.3 Statistical Analyses

After estimating the proportions of different social locations, two logistic regression models were run. First model included main explanatory variables for additive analysis. Second model included only the intersectional location variable with 16 categories for intersectional analysis. A nonparametric bootstrap approach with 1000 replications was used to calculate 95% confidence intervals [30]. I conducted post hoc analysis to calculate probabilities (estimated in an additive scale) which are preferred to odds ratios (OR, estimated in a ratio scale) because ORs are often misinterpreted [4, 24]. Probabilities were reported as percentage probabilities (PP) to aid in interpretation. I conducted complete case analyses (i.e., excluded all missing cases) because the proportion of missing cases was less than one percent, and the missingness did not vary across the outcome. Statistics Canada's [39, 40] benchmarking factors were used as weights to describe the sample. Bonferroni adjustment in pairwise comparisons was made to account for multiple comparison errors. Analyses were conducted in Stata version 17.0 [38].

5.4 Findings

After describing the study participants' characteristics, this section presents the results of outcome patterns of Canadians' concern for self-health due to the COVID-19 pandemic across their four single-axis and 16 intersectional social locations.

Participant characteristics Over three-fifths (60.5%, weighted n = 144,743) of the total participants were 40 years and older, half of them (50.6%, weighted n = 121,107) were women, and over one in three (34.1%, weighted n = 81,436) had a nonsecure income, and two-fifths (39.7%, weighted n = 94,884) lived in small communities where less than 500,000 people lived. The weighted percentages of participants in 16 intersectional locations ranged from 2.8 to 12.5%. *Younger men who had nonsecure income and lived in small communities* represented the lowest intersectional group (weighted n = 6,624). In comparison, *older men with secure income who lived in large communities* belonged to the largest intersectional group (12.5%, weighted n = 29,961).

Participants' concern for self-health: An additive versus an intersectional story Some 15.6% (weighted n = 37,251) of participants expressed extreme concern for their own health due to the COVID-19 pandemic in April 2020.

What does an additive analysis show? An additive analysis (Fig. 2)

shows: (a) participants who were less than 40 years and younger compared to their older counterparts, (b) men compared to women, and (c) participants who had a secure income compared to those who did not, and (d) those who lived in large compared to small communities had lower probabilities of being concerned about their self-health.

No doubt that the outcome estimates for these eight single-category locations are informative, but a descriptive intersectional analysis compared to an additive analysis tells us a far more complex, nuanced, and informative story.

What does an intersectional analysis add to reveal the patterns of the outcome? In the intersectional analysis, the pattern of concerns for self-health across 16 intersections (Fig. 3) tells us that not just any young man but *younger, men, with a secure income, living in small or large communities* are located at the most advantaged location (PP, 5.9–6.6%). In contrast, *younger men, with nonsecure income living in large communities* had 15.8% probability of being concerned; and their probabilities were higher compared to all other intersectional groups of young men and similar to *younger women with nonsecure income, living in either large or small communities (PP, 15.5–16.0%).*

In the spectrum of disadvantaged locations, *older, men, with nonsecure income living in either small or large communities* had over 21% probability of being concerned. However, *the most disadvantaged intersectional groups were older women with nonsecure income living in small or larger communities.* They had at least a 25.0% probability of being concerned about their health. As the dataset did not have information on structures, I turned to the literature to find out how the interacting social systems of power might have contributed to inequalities.

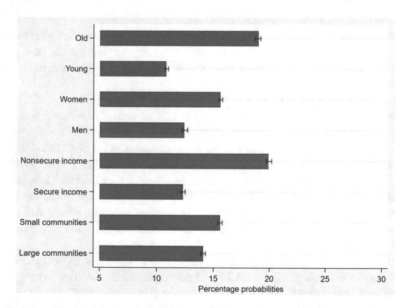

Fig. 2 An additive analysis, showing survey participants' concerns for self-health across their single-category social locations, estimated in percentage probabilities (PP = probabilities × 100). Horizontal lines with spikes at the right edge of each bar show 95% confidence intervals. *Source* Statistics Canada's COVID-19 Impacts Survey, April 2020, N = 239,143

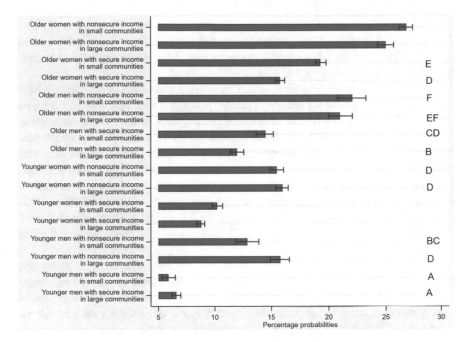

Fig. 3 An intersectional analysis, showing survey participants' concerns for self-health across their different intersectional social locations, estimated in percentage probabilities (PP = probabilities × 100). Horizontal lines with spikes at the right edge of each bar show 95% confidence intervals. Groups sharing the same letters mean that they are similar. *Source* Statistics Canada's COVID-19 Impacts Survey, April 2020, N = 239,143

6 Discussion

Apparently, Canadian survey participants, in navigating the pandemic, might have been *"in the same storm, but not in the same boat"* (Author unknown, cited in [32], p. 91, emphasis added).

Findings from the additive approach indicate that participants who were older, women, poor or lived in smaller communities were more concerned than younger, men, nonpoor, and large community resident counterparts, respectively. As these four categories overlap, this interpretation of the independent effect of one category (e.g., age) on the outcome is conflated with the effects of others (e.g., gender, income).

On the other hand, findings from the descriptive intercategorical intersectional analysis tell a more complex story and warrant us to inquire about interacting structures. For example, why were *older women, with nonsecure income, living in large/small communities* more concerned about their health than other intersectional groups? While the additive analysis paints younger men with a broad brush as a privileged group, the intersectional analysis contradicts this homogenization, leading us to ask: why are *younger men, with nonsecure income, living in large communities* more concerned than their peers with a secure income, living in either large or small communities? To answer these questions, examining the social structures of power (i.e., ageism, sexism, classism, and the geographical divide) in an independent, stand-alone manner is not useful. Instead, examining how these structures interact to exact power might be more insightful.

Although COVID-19 infection fatality rate (IFR) was much higher among older than younger people due to reduced immunity, the fact that IFR was much higher among people living in assisted care facilities compared to their community-living counterparts points to the socio-economic inequities in care [31]. Older people's increased IFR in interaction with ageism, sexism, and the market-oriented economic system of Canada might have made the older women who lack a secure income have a high probability of getting concerned for self-health. Ageism, which allows discrimination against older people and promotes stereotypes against them, limits older adults from keeping jobs or re-entering the labour force, especially those with lower levels of education and training [17]. Thus, the ageist and economic systems reinforce each other.

Consequently, the interaction among the ageist, economic, and sexist systems may explain why older women with nonsecure income were especially concerned for their health. Because of the historical and ongoing sexism, women in Canada face oppression, manifested in the reality that more women than men are unemployed [41]. Compared to men, they are also doing more precarious, part-time jobs and bearing the brunt of unpaid care work [41]. The interaction between ageist and economic systems also explains why older men with nonsecure income have the second-highest probability of experiencing the adverse outcome. Interestingly, the precarity of younger men, with nonsecure income, who were living in larger communities might be due to the interaction between the economic system and their urban location. Although people in rural areas suffer more from poor infrastructure and lack of economic activities [23], COVID-19 hit hard the cities because of lockdown and a sharp decrease in economic activities, especially in hospitality and service industries, as nonessential workers started working from home [20].

This study has several limitations. The dichotomous coding has caused a loss of information. Sensitivity analyses should have been conducted to check how coding decisions affected results. Although both logistic models converged, the additive model was not correctly specified, and the AU-ROC statistics of both models were moderate (see the criteria in [18]). The non-representative online survey also might not have reached different marginalized populations, including the poor, disabled, and homeless, and it did not allow for collecting participants' ethnicity, race, indigeneity, gender and sexual minority status, and disability. The cross-sectional survey allowed only to show outcome patterns at intersections, representing the association. Using mixed methods studies with active community participation might be more appropriate to interrogate the interacting power structures and processes towards achieving an intersectionality project's health equity goals.

Despite these limitations, the intersectional approach allowed (a) conducting nuanced analysis, (b) generating questions regarding causation, and (c) identifying intervenable factors for action. This COVID-19 descriptive intercategorical intersectional analysis is *one of many ways* of doing intersectional analyses. I, however, hope this chapter will pique readers' interest and encourage them to learn more about and pursue quantitative intersectionality research.

7 Discussion Questions

I am curious to know what questions this chapter might have generated for you and your students. For example, can you identify research objectives and questions using intersectional and additive approaches for a topical health problem? In what scenario might an intersectional compared to an additive approach not provide additional insights? What interacting structures, systems, and processes are you most concerned about? How can you ask questions to measure them? In what other ways can you collect data from marginalized populations? Finally, how might you tell an intersectional story to inspire change?

Acknowledgements I am deeply indebted to Greta Bauer, Ph.D., the former Canadian Institutes for Health and Research Sex and Gender Science Chair and currently an Adjunct Professor at the University of Western Ontario, for providing guidance in writing this chapter. I am immensely thankful to the editors and reviewers for their insightful feedback on making this chapter accessible to an interdisciplinary audience.

Funding Post-doctoral research fund of Dr. Greta Bauer's Sex and Gender Science Chair (GSB-171372) of the Canadian Institutes for Health Research.

References

1. Alexander-Floyd, N.G.: Disappearing acts: Reclaiming intersectionality in the social sciences in a post-black feminist era. Fem. Form. **24**(1), 1–25 (2012)
2. Ali, S., Asaria, M., Stranges, S.: COVID-19 and inequality: Are we all in this together? Can. J. Public Health **111**(3), 415–416 (2020)
3. Alliance for Healthier Communities.: Letter to premier ford, deputy premier elliott and Dr. Williams regarding the need to collect and use socio-demographic and race-based data. (2020) https://www.allianceon.org/news/Letter-Premier-Ford-Deputy-Premier-Elliott-and-Dr-Williams-regarding-need-collect-and-use-socio
4. Bauer, G.R.: Incorporating intersectionality theory into population health research methodology: Challenges and the potential to advance health equity. Soc. Sci. Med. **110**, 10–17 (2014)
5. Bauer, G.R., Churchill, S.M., Mahendran, M., Walwyn, C., Lizotte, D., Villa-Rueda, A.A.: Intersectionality in quantitative research: A systematic review of its emergence and applications of theory and methods. SSM—Population Health **14**, 100798 (2021)
6. Bauer, G.R., Scheim, A.I.: Methods for analytic intercategorical intersectionality in quantitative research: Discrimination as a mediator of health inequalities. Soc. Sci. Med. **226**, 236–245 (2019)
7. Berkman, L.F., Kawachi, I.: A historical framework for social epidemiology: Social determinants of population health. In Berkman, L.F., Kawachi, I., Glymour M.M. (Eds.), Social epidemiology (2nd ed.). Oxford University Press (2015)

8. Bowleg, L.: The problem with the phrase women and minorities: Intersectionality—an important theoretical framework for public health. Am. J. Public Health (1971), **102**(7), 1267–1273 (2012)
9. Bowleg, L.: "Once you've blended the cake, you can't take the parts back to the main ingredients": Black gay and bisexual men's descriptions and experiences of intersectionality. Sex Roles **68**(11–12), 754–767 (2013)
10. Bowleg, L., Teti, M., Malebranche, D.J., Tschann, J.M.: "It's an uphill battle everyday": Intersectionality, low-income Black heterosexual men, and implications for HIV prevention research and interventions. Psychology of Men & Masculinities **14**(1), 25–34 (2013)
11. Camus, A.: The Plague (Gilbert, S. Trans.) (1948)
12. Collins, P.H.: Intersectionality as critical social theory. Duke University Press (2019)
13. Collins, P.H., Bilge, S.: Intersectionality. Polity Press (2016)
14. Combahee River Collective: Combahee river collective statement. (1977). https://americanstudies.yale.edu/sites/default/files/files/Keyword%20Coalition_Readings.pdf
15. Crenshaw, K.: Demarginalizing the intersection of race and sex: A black feminist critique of antidiscrimination doctrine, feminist theory, and antiracist politics. Univ. Chic. Leg. Forum **14**, 538–554 (1989)
16. Debczak, M.: Revised guidelines redefine birth years and classifications for Gen X, Millennials, and Generation Z. Mental Floss. (2019, December 6). https://www.mentalfloss.com/article/609811/age-ranges-millennials-and-generation-z
17. Employment and Social Development Canada.: Harassment and sexual violence in the workplace public consultations: What we heard. (2017). https://www.deslibris.ca/ID/10093254
18. Fisk, S.A., Lindström, M., Perez-Vicente, R., Merlo, J.: Understanding the complexity of socioeconomic disparities in smoking prevalence in Sweden: A cross-sectional study applying intersectionality theory. BMJ Open **11**(2), e042323 (2021)
19. Fisk, S.A., Mulinari, S., Wemrell, M., Leckie, G., Perez Vicente, R., Merlo, J.: Chronic obstructive pulmonary disease in Sweden: An intersectional multilevel analysis of individual heterogeneity and discriminatory accuracy. SSM—Population Health **4**, 334–346 (2018)
20. Haider, M., Moranis, S.: Why the pandemic's lingering effects will continue to hurt the hospitality industry. Financ Post. (2021, February 26). https://financialpost.com/real-estate/property-post/why-the-pandemics-lingering-effects-will-continue-to-hurt-the-hospitality-industry
21. Hancock, A.-M.: When multiplication doesn't equal quick addition: Examining intersectionality as a research paradigm. Perspect. Polit. **5**(01) (2007)
22. Hankivsky, O.: Women's health, men's health, and gender and health: Implications of intersectionality. Soc. Sci. Med. **74**(11), 1712–1720 (2012)
23. Infrastructure Canada.: Rural opportunity, national prosperity: An economic development strategy for rural Canada. Infrastruct. Can. (2019) https://ised-isde.canada.ca/site/rural/en/rural-opportunity-national-prosperity-economic-development-strategy-rural-canada
24. Jaccard, J.: Interaction effects in logistic regression. SAGE Publications. (2001)
25. Krieger, N.: Epidemiology and the people's health: Theory and context (Illustrated edition). Oxford University Press. (2011)
26. LaRochelle-Côté, S., Uppal, S.: Differences in the concerns of Canadians with respect to the COVID-19 pandemic (Catalogue no. 45280001; StatCan COVID-19: Data to Insights for a Better Canada). Statistics Canada. (2020)
27. Lofters, A., O'Campo, P.: Differences that matter. In: O'Campo, P., Dunn, J.R. (eds.) Rethinking social epidemiology, pp. 93–109. Springer, Netherlands (2012)
28. Martínez-Mesa, J., González-Chica, D.A., Duquia, R.P., Bonamigo, R.R., Bastos, J.L.: Sampling: How to select participants in my research study? An. Bras. Dermatol. **91**(3), 326–330 (2016)
29. McCall, L.: The complexity of intersectionality. Signs: J. Women Cult. Soc. **30**(3), 1771–1800. (2005)
30. Mooney, C., Duval, R.: Bootstrapping. SAGE Publications, Inc. (1993)
31. Perez-Saez, J., Lauer, S.A., Kaiser, L., Regard, S., Delaporte, E., Guessous, I., Stringhini, S., Azman, A.S., Alioucha, D., Arm-Vernez, I., Bahta, S., Barbolini, J., Baysson, H., Butzberger,

R., Cattani, S., Chappuis, F., Chiovini, A., Collombet, P., Courvoisier, D., Valle, A.Z.: Serology-informed estimates of SARS-CoV-2 infection fatality risk in Geneva. Switzerland. The Lancet Infectious Diseases **21**(4), e69–e70 (2021)

32. Poteat, T.: Navigating the storm: How to apply intersectionality to public health in times of crisis. Am. J. Public Health **111**(1), 91–92 (2021)
33. Rahman, L.: Stata code: Additive and intersectional analyses for self-health concerns during the COVID-19 pandemic in Canada. (2022). https://doi.org/10.17632/jwrdck627d.1
34. Rahman, L., Du Mont, J., O'Campo, P., Einstein, G.: Intersectional inequalities in younger women's experiences of physical intimate partner violence across communities in Bangladesh. Int. J. Equity Health **21**(1), 4 (2022)
35. Ramirez, J.L., Paz Galupo, M.: Multiple minority stress: The role of proximal and distal stress on mental health outcomes among lesbian, gay, and bisexual people of color. J. Gay & Lesbian Ment. Health **23**(2), 145–167 (2019)
36. Rothman, K.J., Greenland, S., Poole, C., Lash, T.L.: Chapter 2: Causation and causal inference. In Rothman, K.J., Greenland, S., Lash T. L. (Eds.), Modern epidemiology (3rd edition). Lippincott Williams & Wilkins. (2012)
37. Spierings, N.: The inclusion of quantitative techniques and diversity in the mainstream of feminist research. Eur. J. Women's Stud. **19**(3), 331–347 (2012)
38. StataCorp.: Stata statistical software: Release 17.0. StataCorp LLC. (2021)
39. Statistics Canada.: Crowdsourcing: Impacts of COVID-19 on Canadians. (2020a, June 3). https://doi.org/10.25318/45250003-eng
40. Statistics Canada.: User guide for the crowdsourcing: Impacts of the COVID-19 on Canadians, public use microdata file. (2020b). https://doi.org/10.25318/45250003-eng
41. World Economic Forum.: Global gender gap report 2021. World Economic Forum. (2021). https://www.weforum.org/reports/global-gender-gap-report-2021/
42. World Health Organization. (2011). *Rio political declaration on social determinants of health*. World Health Organization. https://www.who.int/publications/m/item/rio-political-declaration-on-social-determinants-of-health
43. World Health Organization.: COVID-19 strategy update. (2020). https://www.who.int/publications/m/item/covid-19-strategy-update

Community-Based Participatory Research and Respondent-Driven Sampling: A Statistician's, Community Partner's and Students' Perspectives on a Successful Partnership

M. A. Rotondi, D. Jubinville, S. McConkey, O. Wong, L. Avery, C. Bourgeois, and J. Smylie

Abstract Community-based participatory research fully integrates statistical scientists into the research team, creating dynamic research relationships where both researchers and community partners are educated. Graduate students are also provided opportunities to collaborate with diverse stakeholders and develop skills in knowledge translation. A community-based framework is optimal for studying hard-to-reach populations since community ownership of study processes and results ensures research questions better reflect the community's priorities and needs. Respondent-driven sampling has become increasingly popular as a survey and analysis technique within community-based participatory research due to its ability to recruit hard-to-reach populations more effectively and with less bias than traditional sampling techniques. This chapter focuses on the experiences of the authors

---

M. A. Rotondi (✉) · O. Wong
School of Kinesiology and Health Science, York University, Toronto, ON, Canada
e-mail: mrotondi@yorku.ca

O. Wong
e-mail: owong3@my.yorku.ca

D. Jubinville
Faculty of Health Sciences, Simon Fraser University, Vancouver, BC, Canada
e-mail: danette_jubinville@sfu.ca

S. McConkey · J. Smylie
Dalla Lana School of Public Health, University of Toronto and Well Living House, St. Michael's Hospital, Toronto, ON, Canada
e-mail: stephanie.mcconkey@unityhealth.to

J. Smylie
e-mail: janet.smylie@utoronto.ca

L. Avery
Princess Margaret Cancer Centre, University Health Network, Toronto, ON, Canada
e-mail: lisa.avery@uhnresearch.ca

C. Bourgeois
Seventh Generations Midwives Toronto, Toronto, ON, Canada
e-mail: cbourgeois@sgmt.ca

within community-based research partnerships which incorporate respondent-driven sampling. Several areas of reflection and suggestions for successful community-based research partnerships for statistical scientists and trainees are highlighted through their stories.

Keywords Community-based participatory research · Respondent-driven sampling · Research methods · Indigenous health · Statistics education · Knowledge translation

1   Introduction

Statistical scientists have a variety of opportunities to contribute to interdisciplinary projects. However, community-based participatory research is unique as the statistician's role goes beyond standard or routine statistical consulting and, instead, they are fully integrated into the research team. Community-based participatory research is defined as "an approach to research that involves collective, reflective, and systematic inquiry in which researchers and community stakeholders engage as equal partners in all steps of the research process with the goals of educating, improving practice or bringing about social change" ([26]). In this way, both researchers and community partners are educated as they both offer their unique strengths and insights in this dynamic research relationship [4, p. 2]. In this framework, statisticians will interact and work directly with other researchers, community stakeholders and community members as equal partners where all team members share a voice in decision making and determining study priorities. Moreover, community-based research provides many opportunities for graduate students, including development of skills in advocacy, knowledge translation, and collaboration with diverse stakeholders [16].

This community-based framework, ensuring community ownership of study processes and results, has become an optimal approach for studying the health and well-being of hidden or hard-to-reach communities that experience marginalization (e.g., people who are experiencing homelessness, urban Indigenous communities, or people who inject drugs, etc.). Working in partnership with community partners and stakeholders ensures research questions better reflect the community's priorities and needs. However, one of the fundamental challenges of obtaining valid statistics for these hard-to-reach or hidden populations is the difficulty in gaining access to a representative sample. In the absence of a sampling framework, traditional sampling strategies for these populations, such as targeted sampling or snowball sampling, are affected by various biases due to limitations in recruitment strategies, location, and choice of initial seed participants [11].

Respondent-driven sampling (RDS) is a survey and analysis technique that uses network sampling to obtain asymptotically unbiased estimators of prevalence in these hidden populations [8, 10, 14]. Through adjustments for network size (the number of individuals the participant knows in the population) and homophily (the tendency to recruit individuals with similar characteristics), RDS is able to overcome these biases as RDS samples become representative of the population after several recruitment waves [10, 21]. In RDS, recruitment begins with a group of 'seeds' who are selected from the population. This technique is similar to snowball sampling but restricts the number of recruits (usually around 3), so that after four or five waves potential bias caused by the choice of the initial seeds is reduced. In addition, advanced methodology is used to analyse the data which further controls for the homophily between respondents (the extent to which people with similar traits cluster together) and their selection probability. As an example, a graphical representation of the recruitment process for *Our Health Counts (OHC) Toronto* is included in [19]. More recently, newer estimators for RDS have been developed, including the successive sampling estimator [9] and the homophily configuration graph [6]. These newer estimates go beyond adjusting for a participant's sampling probability and incorporate information about the disease (or other demographic) status of recruiters and their recruits to mitigate the influences of homophily.

Given its ability to penetrate hard-to-reach or hidden populations more effectively and with less bias than traditional sampling strategies, RDS has a breadth of applicability for studying the health of communities experiencing marginalization. Due to such advantages, RDS has exploded in popularity and is increasingly used in community-based public health research. As this sampling technique relies on peer recruitment through community members, this study design naturally lends itself to a community-based framework to ensure research questions are aligned with the community's priorities and needs. Recognizing that the RDS design has significant analytic and statistical complexities, statistical scientists are now commonly asked to be involved in these community-based participatory research partnerships and co-lead the design and analysis of RDS studies.

This chapter is structured in three remaining sections: Michael Rotondi is an Associate Professor of Biostatistics and Quantitative Methods at York University and will share his experiences in community-based research and keys for a successful research partnership. Following this, three graduate students share their experiences as quantitative health researchers involved in community-based participatory research using RDS. Danette Jubinville is a Ph.D. candidate at Simon Fraser University, Faculty of Health Sciences. Stephanie McConkey is a Ph.D. student in Epidemiology and Vanier scholar at the Dalla Lana School of Public Health at the University of Toronto. Octavia Wong is a Ph.D. candidate at York University's School of Kinesiology and Health Science. Finally, our community partner, Cheryllee Bourgeois, an Exemption Metis Midwife, at Seventh Generation Midwives Toronto, shares her experiences working with statisticians on the *OHC Toronto* project. Through the sharing of these stories, we highlight several areas of reflection and suggestions for successful community-based research partnerships for statistical scientists and trainees.

2   A Statistical Lead's Keys for a Successful Partnership

Over the past ten years, I have worked on a variety of community-based partnerships using RDS, including *Change the Cycle* [24] and *OHC Toronto* [7]. *Change the Cycle* [24] was a community-based intervention targeted towards people who inject drugs, which was designed to reduce the risk of initiating injection drug use in non-injectors. Based on social learning theory and adapted from a similar program in the United Kingdom [12], this program aimed to reduce injection drug use. Under the fundamental premise of Indigenous community ownership and self-determination, the *OHC Toronto* study successfully recruited a sample of nearly 1000 Indigenous people who live in Toronto and represents the largest, single source of urban Indigenous health information in Canada. From these experiences, I have the following recommendations for statisticians who are interested in undertaking, or are actively engaged in, community-based participatory research, both within the context of RDS studies and community-engaged participatory research more broadly.

2.1   Commit to Furthering Your Understanding of the Community and Their Experiences

As with any collaborative partnerships, statisticians should develop a fundamental understanding of the research problem. This will allow for more detailed and nuanced discussions of aspects of modelling and will also be an opportunity to gain new knowledge and perspective. While this may apply to statistical consulting in general, in my experience this has an even greater significance in the context of community-based participatory research and RDS studies. In addition to contributing to the research leadership team, formulating research questions, and generating new knowledge in support of the community, it is important for the statistician to be sensitive and understand the experiences of community members.

2.2   Seek Support of a "Subject Translator"

Statisticians are often responsible for translating available data into useful information, but there are cases where statisticians may struggle to convey the results of some advanced statistical models or techniques. In these cases, it is often helpful to have a "subject translator" available, who can understand both the advanced models and help convey their meaning to stakeholders. In our experience in health research, having an epidemiologist or medical professional is often helpful as these individuals understand both the statistical relevance and potential usefulness and impact for our community partners and stakeholders.

2.3 Respectful Engagement of Partners

We must also recognize that a true partnership combines the strengths of all team members to reach the project's goals. While the statistician is often the sole statistical expert and may have extensive training and experience in the area, they must recognize that all team members bring their own unique perspectives and strengths to the project and be respectful of all viewpoints and contributions. Similarly, it is important to recognize that something which may appear optimal from their perspective, may not be appropriate for the community. Through discussion and respectful engagement, the research team can reach an appropriate solution balancing statistical rigour with the community's needs.

2.4 Understand the "Bigger Picture"

As statisticians, we may not fully understand the importance of how these analyses and statistical results fit the goals and objectives of our community partners and stakeholders. In the context of urban Indigenous health research, there is often very little quantitative data available, thus while we may strive for a perfect and precise answer, in some cases even a less precise answer may provide immense value to our stakeholders. As an example, in the *OHC Toronto* study, stakeholders were certain that the Canadian Census has vastly undercounted the size of the Indigenous Community in Toronto. This had significant policy implications for the delivery of social and health services for this urban Indigenous community. Through ongoing modelling discussions and presentations to the community partners, we showed that under a conservative model, the Canadian Census underestimated the size of the Indigenous community in Toronto by a factor of two to four [19]. From a statistical perspective, we were concerned that this estimate was not sufficiently precise for use, but this provided sufficient information for our community partners and stakeholders to further advocate for resources to support the needs of the Indigenous community in Toronto. In this way, simply showing that the official census count was incorrect was far more important than the actual range.

2.5 Recognize When Methodological Details Are Not Needed

Similar to the importance of statisticians understanding the bigger picture, it is important to recognize that some stakeholders are not always interested in aspects such as modelling details or statistical assumptions. In the *Change the Cycle* project, analyses were complicated by missing data at a second longitudinal time point. Given the lack of a clear methodological approach to resolve this problem, partners and stakeholders were most interested in whether or not their intervention worked but were unable to make this determination. While they recognized that this research question now required advanced techniques that were not currently developed, many team members were less interested in aspects such as simulation details, and simply needed a clear answer to frame the future direction of this research and social program. As statisticians, we are often very excited and enthusiastic about novel statistical ideas, but in some cases, it is important to be responsive to stakeholders' needs and ensure that communication is clear and concise, without added jargon or technical details.

2.6   The Science of Uncertainty

Perhaps the most important aspect of communicating statistics and working with stakeholders is the translation of uncertainty. While not strictly valid, some statistical concepts such as 95% confidence intervals can often be translated using simpler language, such as 'plausible range' of the estimate or similar notions, but given the relative novelty of RDS many statistical challenges remain unaddressed. For example, over the course of my work in urban Indigenous health research, analytic strategies have evolved as more methodological work is performed and biostatistical researchers develop a more advanced understanding of the intricacies of multivariable analyses of RDS data. This has led not only to changes in our approaches, but potentially different results. In these cases, it is most helpful to be transparent and emphasize that as techniques evolve, answers may change. Specifically, the methods that were used to analyze earlier databases, may not be the same as those that are used now. A summary of these tips is presented in Table 1.

Table 1   A Statistician's Tips for a Successful Partnership

Tips
• Improve your understanding of the community and research priorities
• Seek support of an additional translator, if possible
• Ensure respectful engagement of study partners
• Understand how analyses fit and further your partners' objectives
• Recognize when methodological details are not required
• Understand and convey uncertainty

3 Students' Perspectives on Working with Indigenous Community Research Partners and RDS in a Quantitative Discipline

3.1 Indigenous Reproductive Justice and Community Wellness, by Danette Jubinville

My first experience with quantitative research was for my master's thesis, *Resources for Indigenous reproductive justice and wellness in Toronto: A respondent-driven sampling study,* which was nested within the *OHC Toronto* study [15]. My research was a secondary analysis that explored the relationship between land and wellness for women, two-spirit, trans, and gender diverse people of childbearing age. This research stemmed from community priorities as well as my own research interests, which have been shaped by my personal experiences as an urban Cree and Anishinaabe mother and doula living in Vancouver, BC.

Coming from a background in Indigenous studies and qualitative methods, quantitative research involved a steep learning curve. Initially, I had to confront my knowing that statistics have been used to further colonial agendas of dispossession and elimination of Indigenous peoples. Having mentorship from Dr. Janet Smylie, as well as theoretical guidance from Maggie Walter and Chris Andersen's [27] *Indigenous Statistics* helped me to understand that quantitative methods can and should be used to serve Indigenous agendas, and our ancestral ways of knowing demonstrate they always have.

To centre Indigenous knowledge and theories in my statistical model I used a Directed Acyclic Graph (DAG). DAGs are becoming increasingly common in statistical science, as they offer a straightforward method for researchers to identify which variables require conditioning for control of confounding, while providing transparency around the theories and assumptions that undergird an analysis [25]. My model was large (19 covariates) and given that RDS methods require large sample sizes due to large design effects, there were some challenges related to power and precision for my predetermined sample size. Additionally, although RDS helped to capture data for community members that have too often been left out of health research, including Métis, trans, and gender diverse people, these populations were still underrepresented in my study. This limitation required mindfulness around generalizing my conclusions to these community members. For future researchers using RDS, I would consider designing a DAG prior to data collection to help determine sample size and prevent possible limitations due to power, precision, and adequate representation.

3.2 The Determinants of Indigenous Peoples' Health as Predictors for Diabetes and Unmet Health Needs, by Stephanie McConkey

For my thesis in the Master of Science in Epidemiology and Biostatistics program at Western University, my research was nested in the larger multi-year *OHC* project; created and developed by Dr. Janet Smylie and multiple Indigenous community partners in Ontario cities. My thesis project was titled "The Indigenous determinants of health as predictors for diabetes diagnosis and unmet health needs in an urban Indigenous population: a respondent-driven sampling study in Toronto, Ontario" [17] and took me three years to complete because I spent the first year developing a working research relationship with the Indigenous community partners, Seventh Generation Midwives Toronto. Taking an Indigenous-partnered approach, we co-developed a research question and data analysis plan that focused on the community priorities. The statistical model for this research was built using an Indigenous conceptual framework of the determinants of Indigenous peoples' health [18] to understand the unmet health needs among Indigenous peoples living in urban and related homelands and the distal, intermediate, and proximal determinants of diabetes. Two separate analyses were performed for each of the outcomes—unmet health needs and diabetes. The distal, intermediate, and proximal factors were treated as explanatory variables and were blocked together in the logistic regression models based on whether the variable of interest was a distal, intermediate or proximal determinant of Indigenous peoples' health (see Figs. 1 and 2). Using Indigenous understandings and concepts of health to build statistical models is one example of the application of Indigenous research methodologies in quantitative research.

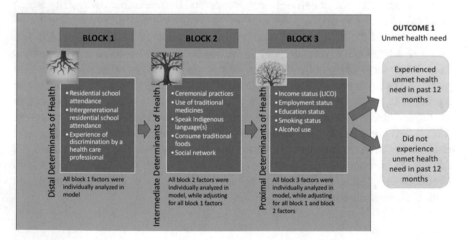

Fig. 1 Distal, intermediate and proximal determinants of unmet health needs among Indigenous peoples living in urban and related homelands (reprinted from [17] with permission)

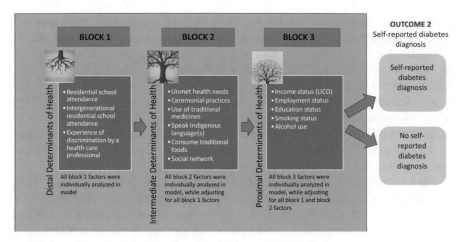

Fig. 2 Distal, intermediate and proximal determinants of diabetes among Indigenous peoples living in urban and related homelands (reprinted from [17] with permission)

In my experience as a First Nations epidemiologist, western-based quantitative research disciplines inadequately respect, understand and incorporate Indigenous research methodologies. Therefore, I strive to ensure the Indigenous community partners are involved in all aspects of our research—from developing our research question and objectives to all knowledge translation activities. Every statistic holds a story; therefore, it is extremely important to tell these stories using Indigenous paradigms and ways of knowing. This teaching has driven my passion to develop statistical models using Indigenous conceptual and theoretical frameworks (i.e., the determinants of Indigenous peoples' health), as well as always ensure the community is included in telling these stories by taking an Indigenous-partnered approach to research, an important methodology that is often left undone in quantitative disciplines.

3.3 Risk and Protective Factors for Major Depressive Disorder in Urban Indigenous Communities, by Octavia Wong

As an allied statistician, I have been working on a community-based partnership using RDS, the *OHC* studies as a part of my graduate studies. My master's thesis, "Identification of risk and protective factors: A study of major depressive disorder among Indigenous adults in Toronto" [28], was nested within the *OHC Toronto* study. This research was focused on identifying factors that play a role in the rates of major depressive disorder within the Indigenous community in Toronto. Throughout the process of writing my thesis, with the guidance of the leadership team and community partners, I began to understand the importance of using a strength-based approach for community-based research. Oftentimes, when we examine factors that may be related to adverse health outcomes, we focus on the factors within the community that are lacking or detrimental. This deficit-based framework can lead to further stigmatization or marginalization [13]. However, a strength-based approach is community-driven, focusing instead on all the strengths of the community that can be built upon. As such, an advantage of using a strength-based approach in community-based research is how well it lends itself to examining factors that are more relevant to the community and contribute to their resilience, such as access to traditional medicines and ceremonies in the Indigenous community.

Most recently, for my PhD dissertation, I have been working with the *OHC* community partners to develop and validate meta-analysis techniques to summarize RDS data across *OHC* study sites. Currently, results from each study site's RDS data alone do not have enough precision to make strong conclusions. The successful development of an appropriate meta-analysis technique for RDS studies would improve the methodological rigor of RDS analyses. In some cases, our small sample sizes and large design effects due to RDS contribute to low levels of precision, which limits the strength of our conclusions. With the increased power and precision from these meta-analysis techniques, we can more accurately examine outcomes prioritized by our Indigenous community partners while ensuring that their needs, concepts, and culture are represented in every aspect of data analysis and interpretation.

4   A Community Partner's Perspective on the Importance of Working with Statisticians

Indigenous midwives traditionally hold the longitudinal health data of the communities they serve—in layperson's terms, they are the ones who know who is related to whom, which families are more likely to have a breech baby or twins, and what health issues are more common in which areas. I take this role very seriously in my midwifery work and see research projects as an extension of this cultural responsibility.

Indigenous people deserve the right to services, education and good quality data that reflect their Nationhood, culture, and knowledge. For this reason alone, it is critical that researchers and statisticians involved in Indigenous data analysis understand that within each data point exists the experiences, lived realities and medicine of Indigenous peoples.

As demonstrated above, when research includes Indigenous community partners in the development of meaningful statistical analysis models and methods, it advances the potential for transformational and impactful social change at a community level as well as in policy and institutional practice. For example, *OHC* projects have provided Indigenous organizations with data to support what they already knew about

Table 2 Reflections on indigenous community partnerships

Reflections on Indigenous Community Partnerships
• Every statistic represents a lived experience
• Accurate data is vital for Indigenous communities community to achieve their goals
• Strength-based analyses and sharing of study results can help dispel stereotypes and strengthen the community
• Indigenous community leadership and participation in applied statistical projects leads to high quality research and advancement of Indigenous self-determination

community size, strengths, needs, and health disparities. *OHC* datasets also represent a powerful policy tool with which Indigenous community leaders can demonstrate and get policy uptake regarding vast underfunding and the need for Indigenous specific services. This benefits the researcher, leading to papers and presentations, and the Indigenous community, by producing and disseminating good quality and reliable data, which remains a challenge in Indigenous health research. Most importantly, however, the practice is vital to reclaiming Indigenous self-determination in health, wellbeing and community care. Together, we can work towards addressing inequities and improving the health of our community. A summary of these reflections is included in Table 2.

5   Future Directions

The opportunity to engage groups traditionally excluded from social sciences and health research using RDS is an exciting prospect. The COVID-19 pandemic has caused a shift toward more flexible data collection strategies including telephone, virtual meeting platforms and online surveys. Coupled with social media which offers virtual spaces for those with similar lived experiences, RDS offers an opportunity to allow researchers to engage with individuals from socially diverse communities and reduce recruitment costs. However, there are also challenges with these novel approaches, including verifying membership in specific populations, accessing private social media platforms and controlling for duplicate responses. The key to overcoming these challenges is a true partnership between community members, researchers and relevant stakeholders. As these projects have shown, when community members are active research participants, and when academic researchers are dedicated to serving a community, it becomes easier to involve members of the community in the recruitment effort, overcome challenges, and ensure study success.

As statisticians and quantitative researchers, we have unique opportunities that are not available to other researchers. Specifically, our interdisciplinary nature allows us to participate in a variety of projects and contribute to new knowledge in many areas. However, given the statistician's role in interpreting the data and information at hand, we also have a responsibility to act ethically. To this end, the Statistical Society of Canada [23], American Statistical Association [3] and Royal Statistical Society [20] have all produced guidelines for the ethical practice of statistics and data science. In addition to the opportunities for personal growth, reflection and learning, perhaps the most rewarding aspect of this working within a community-based participatory research framework is the opportunity to support actionable information for stakeholders, partners and community members. Although there may be some challenges in initiating and maintaining community relationships, a respectful attitude, self-awareness and sincere desire to serve the community can lead to a successful and fulfilling project.

References

1. Avery, L., Nooshin, R., McKnight, C., Firestone, M., Smylie, J., Rotondi, M.: Unweighted regression models perform better than weighted regression techniques for respondent-driven sampling data: results from a simulation study. BMC Med. Res. Methodol. **19**(1), 202 (2019). https://doi.org/10.1186/s12874-019-0842-5
2. Bartlett, D.J., McCoy, S.W., Chiarello, L.A., Avery, L., Galuppi, B.: A collaborative approach to decision making through developmental monitoring to provide individualized services for children with cerebral palsy. Phys. Ther. **98**(10), 865–875 (2018). https://doi.org/10.1093/PTJ/PZY081
3. Committee on Professional Ethics of the American Statistical Association.: Ethical guidelines for statistical practice (2018). https://www.amstat.org/asa/files/pdfs/EthicalGuidelines.pdf
4. Coughlin, S., Smith, S., Fernandez, M.: Overview of community-based participatory research. In: Handbook of Community-Based Participatory Research. Oxford University Press (2017). https://doi.org/10.1093/acprof:oso/9780190652234.001.0001/acprof-978 0190652234-chapter-1
5. Fayed, N., Avery, L., Davis, A.M., Streiner, D.L., Ferro, M., Rosenbaum, P., Cunningham, C., Lach, L., Boyle, M., Ronen, G.M.: Parent proxy discrepancy groups of quality of life in childhood epilepsy. Value in Health **22**(7), 822–828 (2019). https://doi.org/10.1016/j.jval.2019.01.019
6. Fellows, I.E.: Respondent-driven sampling and the homophily configuration graph. Stat. Med. **38**(1), 131–150 (2019). https://doi.org/10.1002/sim.7973
7. Firestone, M., Maddox, R., O'Brien, K., Xavier, C., Wolfe, S., Smylie, J.: Our Health Counts—Project Overview & Methods. Well Living House (2018). http://www.welllivinghouse.com/wp-content/uploads/2019/10/Project-Overview-Methods-OHC-Toronto.pdf
8. Gile, K.J., Handcock, M.S.: Respondent-driven sampling: an assessment of current methodology. Sociol. Methodol. **40**(1), 285–327 (2012). https://doi.org/10.1111/j.1467-9531.2010.01223.x
9. Gile, K.J.: Improved inference for respondent-driven sampling with application to HIV prevalence estimation. J. Am. Stat. Assoc. 106(493), 135–146 (2011). https://www.jstor.org/stable/41415539
10. Heckathorn, D.D.: Respondent-driven sampling: a new approach to the study of hidden populations. Soc. Probl. **44**(2), 174–199 (1997). https://doi.org/10.2307/3096941
11. Heckathorn, D.D.: Respondent-driven sampling II: deriving valid population estimates from chain-referral samples of hidden populations. Soc. Probl. **59**(1), 11–34 (2002). https://doi.org/10.1525/sp.2002.49.1.11
12. Hunt, N., Stillwell, G., Taylor, C., Griffiths, P.: Evaluation of a brief intervention to prevent initiation into injecting. Drugs: Educ. Prev. Policy 5(2), 185–194 (1998). https://doi.org/10.3109/09687639809006684

13. Hyett, S., Gabel, C., Marjerrison, S., Schwartz, L.: Deficit-based Indigenous Hhealth research and the stereotyping of Indigenous peoples. Can. J. Bioeth. Revue Canadienne de Bioéthique **2**(2), 102–109 (2019). https://doi.org/10.7202/1065690ar

14. Johnston, L.G., Hakim, A.J., Dittrich, S., Burnett, J., Kim, E., White, R.G.: A systematic review of published respondent-driven sampling surveys collecting behavioral biologic data. AIDS Behav. **20**(8), 1754–1776 (2016). https://doi.org/10.1007/s10461-016-1346-5

15. Jubinville, D.: Resources to support Indigenous reproductive health and justice in Toronto: a respondent-driven sampling study. Master's thesis, Simon Fraser University, Simon Fraser University: Summit—Institutional Repository (2018)

16. Khobzi, N., Flicker, S.: Lessons learned from undertaking community-based participatory research dissertations: the trials and triumphs of two junior health scholars. Prog. Commun. Health Partnersh. Res. Educ. Action **4**(4), 347–356 (2010). https://doi.org/10.1353/cpr.2010. 0019. PMID: 21169713

17. McConkey, S.: The Indigenous determinants of health as predictors for diabetes and unmet health needs among urban Indigenous people: a respondent-driven sampling study in Toronto, Ontario. Master's thesis, The University of Western Ontario. Western Graduate & Postdoctoral Studies: Electronic Thesis and Dissertation Repository (2018)

18. Reading, C.: Structural determinants of Aboriginal peoples' health. In: Greenwood, M., de Leeuw, S., Lindsay, N., Reading, C. (eds.) Determinants of Indigenous peoples' health in Canada, pp. 3–15. Canadian Scholars' Press (2015)

19. Rotondi, M.A., O'Campo, P., O'Brien, K., Firestone, M., Wolfe, S.H., Bourgeois, C., Smylie, J.K.: Our Health Counts Toronto: using respondent-driven sampling to unmask census under-counts of an urban Indigenous population in Toronto. Can. BMJ Open **7**, e018936 (2017). https://doi.org/10.1136/bmjopen-2017-018936

20. Royal Statistical Society & Institute and Faculty of Actuaries.: A guide for ethical data science (2019). https://www.actuaries.org.uk/system/files/field/document/An%20Ethi cal%20Charter%20for%20Date%20Science%20WEB%20FINAL.PDF

21. Salganik, M.J., Heckathorn, D.D.: Sampling and estimation in hidden populations using respondent-driven sampling. Sociol. Methodol. **34**(1), 193–240 (2004). https://doi.org/10. 1111/j.0081-1750.2004.00152.x

22. Smylie, J., Firestone, M., Cochran, L., Prince, C., Maracle, S., Morley, M., Mayo, S., Spiller, T., McPherson, B.: Our Health Counts: Urban Aboriginal health database research project—community report. Well Living House (2011). http://www.ourhealthcounts.ca/images/PDF/ OHC-Report-Hamilton-ON.pdf

23. Statistical Society of Canada.: Code of ethical statistical practice (2004). https://ssc.ca/sites/ default/files/data/Members/public/Accreditation/ethics_e.pdf

24. Strike, C., Rotondi, M., Kolla, G., Roy, É., Rotondi, N., Rudzinski, K., Balian, R., Guimond, T., Penn, R., Silver, R., Millson, M., Sirois, K., Altenberg, J., Hunt, N.: Interrupting the social processes linked with initiation of injection drug use: Results from a pilot study. Drug Alcohol Depend **137**, 48–54 (2014). https://doi.org/10.1016/j.drugalcdep.2014.01.004

25. Tennant, P.W.G., Murray, E.J., Arnold, K.F., Berrie, L., Fox, M.P., Gadd, S.C., Harrison, W.J., Keeble, C., Ranker, L.R., Textor, J., Tomova, G.D., Gilthorpe, M.S., Ellison, G.T.H.: Use of directed acyclic graphs (DAGs) to identify confounders in applied health research: review and recommendations. Int. J. Epidemiol **50**(2), 620–632 (2021). https://doi.org/10.1093/ije/ dyaa213

26. Tremblay, M.C., Martin, D.H., McComber, A.M., McGregor, A., Macaulay, A.: Understanding community-based participatory research through a social movement framework: a case study of the Kahnawake Schools Diabetes Prevention Project. BMC Public Health **18**(1), 487 (2018). https://doi.org/10.1186/s12889-018-5412-y

27. Walter, M., Andersen, C.: Indigenous Statistics: A Quantitative Research Methodology. Left Coast Press (2013)

28. Wong, O.: Identification of risk and protective factors: A study of major depressive disorder among Indigenous adults in Toronto. Master's thesis, York University. YorkSpace—Institu-tional Repository: Electronic Theses and Dissertations (2019)

Operationalizing Learning Processes Through Learning Analytics

Alexandra Patzak and Jovita Vytasek

Abstract Recent advances in the use of learning technologies for both in-person and distance education has enabled the collection of detailed data on learners' access and use of resources at unprecedented levels. Simultaneously, the growth of technology use in the classroom has brought forward increased interest in the analysis and use of learner data. The field of learning analytics leverages these data with the aim to enhance understanding about and improve learning processes. The Society of Learning Analytics Research defines learning analytics as "the measurement, collection, analysis and reporting of data about learners and their context" (LAK in 1st international conference on learning analytics and knowledge, Banff, AB, Canada, 2011; SOLAR in What is learning analytics? 2021). This chapter provides an overview of leveraging learning analytics to enhance understanding and provide feedback about learning processes. We aim to offer insights into types of data used to generate learning analytics, use the research on procrastination as an example for interpreting and operationalizing learning processes through data and analytics, and offer recommendations for generating feedback about these data. We aim to offer a starting point for utilizing learning analytics and convey their potential to aid learning and pedagogical choices.

1 Types of Learner Data

One of the unique features of learning analytics is its focus on learners' actions in the *processes* of learning, not just learning outcomes. Learning analytics therefore utilizes data that represents these activities, such as traces of learners' online activity. Trace data generates pieces of information about *when* and *what* a learner interacts

A. Patzak (✉)
College of Education and Human Development, George Mason University, Fairfax, VA, USA
e-mail: apatzak@gmu.edu

J. Vytasek
Faculty of Educational Support and Development, Kwantlen Polytechnic University, Surrey, BC, Canada
e-mail: jovita.vytasek@kpu.ca

D. G. Woolford et al. (eds.), *Applied Data Science*, Studies in Big Data 125,
https://doi.org/10.1007/978-3-031-29937-7_6

with while learning, that researchers and educators can then use to infer *how* a learner engages with a task. Recognizing that these data are only part of a larger picture of interaction, data provides insight into the learning process—for learning analytic designers and/or educators, who may often be the same individual.

Data available about the learner can be classified into 4 categories: activity data, artifact data, association data, and archival data (see [80] for an elaborated description). Activity data encompasses traces of what learners did, such as submitting a quiz. Artifact data are the things that learners created in or summitted to the system, such as the answers to quiz questions. Association data are trace-data recording engagement with the system. This could be between individuals (learner-educator, learners in a class etc.) or between individual(s) and an artifact in the system (i.e., learner opening a course module). Archival data describe any information about the learner that was created during the learning activity, such as demographic information, survey responses, prior course registration, grades (either prior or current performance), etc.

Recording learning processes and activities often generates large quantities of data [37]. A single action can create multiple types of data. For example, the short action of a learner replying to a peer's post in a discussion forum generates multiple types of data, including: (i) an artifact of the post text; (ii) activity trace of learner *A* posting a reply to learner *B* and time of posting; and (iii) association data linking learner *A* to discussion thread *Z*, post text, and learner *A* to learner *B* [80]. Learning environment data can also be added to describe the context of the learning activity (e.g., instructional guidance given for posting in the discussion, learning outcomes, curriculum, instructional design, etc.). This can be valuable metadata, especially when multiple courses are being studied. However, we need to keep in mind that although collecting all these types of data are possible, learning analytics designers and users are often limited by the types of data and level of specificity available from learning management systems (LMS) [70].

To more fully capture learning, it is important to also consider data outside of computer-based systems. Otherwise, we run the risk of allowing the available data to determine what we research [80]. Multimodal Learning Analytics emerged as a field to explore how to research and integrate different types of data not typically captured through computer-mediated systems (see [46] for details). Multimodal Learning Analytics utilizes multimodal data, defined as data that "originates from different data channels which are subjective and/or objective" [25]. Subjective multimodal data include, for example, self-reported intentions to learn or beliefs about learning, while objective multimodal data are often comprised of physiological responses, such as heart rate, galvanic skin response and electroencephalogram (EEG) readings or movement measures and tracking learner gaze, gesture, posture, and so forth (for a detailed list see [5]). Multimodal data can be used to combine varied sources of data to model dynamic multifarious constructs, such as engagement [84], empathy [24], or collaboration [2, 66]. For example, eye-movement data, facial recognition data, and learner actions have been combined to indicate cognitive demand, boredom, or confusion [13, 16]. Multimodal data can provide a more complete picture of learning processes but also requires research about the use of diverse and complex data sources [25, 57].

An important question for learning analytics is how to select, aggregate and report data to benefit learning, teaching, and decision-making about future activities [3, 4]. While digital systems can record with time stamped accuracy some of learners' preferences and choices about learning (e.g., modules opened, resources downloaded, quizzes completed, and assignments submitted to the system, [33, 76]) determining what inferences to make from the sequences and series of actions presents a complex question. *What do those traces mean?* In the next section we will explore this question using academic procrastination as an example.

2 Using Learner Data to Operationalize Procrastination

Time stamps are one of the most prolific aspects of data collected and used in analysis for learning analytics [34, 54]. Timing of learning activities may be ubiquitous data, but the interpretations made are complex and multifaceted [34]. For example, assignment submission time is used to calculate late or near deadline submission. This is often described as a behavioral manifestation of procrastination [1, 21, 55]. From the time stamp when learners submitted their assignments, we can identify when assignments were submitted, which learners submitted assignments after the deadline, and how late assignments were submitted. Does it matter if a learner submits an assignment one minute versus four hours late? Is this an indication of more extreme procrastination? How do we identify when students are procrastinating? Before heading to the trace data logs, it is important to draw on theories and empirical research to guide us in answering these questions. Then we must consider what data might best answer these questions, and if aspects of the learning activity are not recorded, what they might mean for interpreting the data.

As a first step, we need to create a conceptual definition of procrastination and review the literature on approaches to measuring procrastination. This builds the foundation for an operational definition of procrastination based on trace data. Operational definitions are theory and research driven guides on how to interpret traces of behavior (e.g., [65]).

Procrastination is described as the intentional delay of tasks, despite expected negative consequences due to the delay [60, 63]. Researchers have explored why learners procrastinate through cognitive, emotional, and motivational lenses [49]. Temporal motivation theory [63, 64] describes that learners procrastinate when they judge low prospects for success, low enjoyment while engaging in a task, or they perceive a delay between task engagement and expected outcomes. This means, learners' self-efficacy beliefs, outcome expectations, and emotions experienced in relation to task characteristics affect their decision to procrastinate. Grunschel et al. [19] interviewed university learners about reasons to procrastinate. Learners described task characteristics (e.g., task difficulty), lecturer characteristics (e.g., unrealistic expectations), and a lack of competencies and counterproductive self-regulated learning as primary reasons. Overall, procrastination appears to be an individualized, multi-faceted construct that manifests as the tendency to delay actions, referred to in the literature as dilatory behavior.

Dilatory behavior alone is not sufficient evidence for procrastination. Besides simple forgetfulness, this behavior could also be the result of strategic delaying or prioritization of tasks, which are considered productive time management strategies [9, 30, 38]. These strategies involve strategic planning of the order in which tasks are to be completed, based on cost benefit analyses and available resources. Strategic delay of one task might allow for completion of other more pressing tasks. This might be particularly important for individuals who are navigating multiple obligations and juggling tasks with overlapping deadlines.

To understand why learners choose to delay learning, activity traces of learners' assignment submission time need to be complemented with additional information [48] addresses this issue by developing a measure to trace how learners learn online. In a web application, learners could make productive or counterproductive choices about learning, for example, procrastinating by browsing online to delay work on the learning task. Learners rated how conducive or hindering they expect each of those choices to be before engaging in the learning activity and explained why they made specific choices upon task completion. Learners justified spending time on those webpages to "take a break" or "ease my mind from anxiety". In this controlled laboratory setting, learners were only working on one learning task at a time and their options on how to spend their time was limited to choices in the web application. This allowed for an operationalization of procrastination. Yet, it only approximates a realistic learning situation. As they study, learners could be multi-tasking on- and off-line. This requires triangulation of data types from multiple sources beyond time records (behavioral engagement) to also include proxies for emotional engagement which is crucial for understanding learners' choices about studying.

2.1 Combining Data Sources

Recent developments in learning analytics have complemented readily available trace data recording behavioral and cognitive engagement with data about emotional engagement [50–52]. For example, in a series of studies, Tempelaar et al. [86] explored the use of activity data recording learners' time-on-task, supplemented with survey data from traditional self-report surveys about learners' emotions. They found strong interrelations between emotions and how learners choose to spend their time in the online learning environment. Boredom was particularly strongly and negatively correlated with time spent on a learning task. This finding has important implications for learning analytics feedback. If a bored learner is lagging behind in the course, feedback recommending strategies to combat procrastination, or catch-up may not help the learner regulate their boredom and will likely not be effective. This effect would be overlooked when utilizing only measurements of time to infer procrastination or engagement. Time-on-task alone does not afford inferences about why learners choose to engage in a task and falls short in explaining learners' cognitive or emotional engagement [70]. Venturing beyond readily available trace data can provide a more complete picture about the learner, learning activity and enables tailoring learning analytics to the needs of individual learners. While this research has advanced the field, future work is needed to overcome limitations of using self-report surveys to measure learners' engagement. Self-reports suffer from misreporting due to misremembering of information and social desirability [75]. These measures are also limited to a general one-time view of engagement, often related to the overall course a learner is enrolled in. This approach simplifies the complex, multi-faceted structure of engagement and overlooks that engagement changes as learners learn [28, 67, 71].

Trace data are more fine-grained than self-report surveys and can record how learning unfolds [75, 76]. However, it is important to keep in mind that trace data are not a complete representation of the learning activity. Many educational and psychological constructs involve unobservable components, such as motivation, emotions, beliefs, prior experiences, and so on. Trace data used are thus merely proxies of the construct we aim to measure. Drawing on multiple data types is pivotal to capture the complexity of learning processes. Simply looking at time on task and productivity does not provide enough information to understand the learning process taking place. These are important considerations for those designing and utilizing learning analytics to support learners. Intervening or providing the wrong support or feedback could undermine or negatively redirect the learning process.

3 Using Learner Data to Generate Feedback

As data are used to make inferences about learning processes it is important to consider how this information is communicated to those who will be interpreting and using this information. This is particularly relevant when generating feedback to learners, as information about them is presented to them. Justifying links between learner actions and how we conceptualize learning processes and products is a critical component in the development of meaningful learning analytics feedback [79]. The connection to research and theory is fundamental in guiding the feedback process. Researchers and learners agree that effective feedback should be grounded in research [7, 43], tailored to the recipient [7, 12, 20] and actionable [7, 20]. (For a more detailed review of providing effective feedback see [20] or [78]).

Trace data are records of learning processes, which sets the stage for actionable recommendations, generated as learning analytics interventions or feedback [79, 82]. Sequences of behaviors can indicate individual learners' progress and how learners tackle tasks differently [74]. This information can be used to trigger individualized and actionable feedback. For educators, learning analytics can be used as a pedagogical tool to tailor educational activities and materials to learner needs and preferences [82]. Educators can use learner data to evaluate and reflect on the impact of their learning design (or specific instructional approach) on learners and learning outcomes (e.g., [40]), diagnose challenges individual learners or groups are facing (e.g., [68]), or personalize learning activities to learners (e.g., adaptive learning tools and recommender systems, see [27]).

Although learning analytics can offer new avenues for teaching and engaging with learners, their uptake depends on educators' access and opportunities to leverage them [80]. Barriers such as technological limitations, lack of transparency or clarity in the sources and interpretation of data, understanding about the vast utility of learning analytics, or time may hinder some educators from implementing learning analytics. While resources to develop this level of user expertise are available, the extent to which those opportunities are taken up is unclear. Research is needed to explore and develop suitable supports for the co-design, integration, and implementation of learning analytics theory in practice [41, 56, 82].

For learners, feedback can be offered to promote reflection on progress and processes (e.g., mirroring) or provide recommendations for future activities (e.g., learner-facing dashboards; for a review see [6]. Data-driven learning analytics feedback may also be generated in-situ, utilizing trace data to generate prompts at critical moments during a learning session [79]. For example, a learner who seems to be on the wrong track of solving a problem could receive feedback encouraging them to consider a different approach.

While there is great promise and potential for using feedback generated from learning analytics to support learning, personalizing feedback so that it is meaningful and actionable has been a challenge [17, 26] This is a multifaceted issue addressing, *what* data are used, *why* it is being collected and *how* analytics-based feedback can facilitate learning. Caution has been raised in the analytics community that by tracing, measuring, and visualizing certain learning activities, we are placing a spotlight on these actions and run the risk of overemphasizing only those activities which we are able to measure and present using analytics [14]. Learners may overlook other valuable learning activities which are not measured. They often require support with contextualizing, understanding, and acting on learning analytics information provided [83]. The selection and presentation of data, and its role in the learning environment are important considerations which should be clearly conveyed to the learner [82].

Recent research indicates that learners are not satisfied with current approaches to designing and presenting analytics feedback. Learners feel positioned as the passive recipients of data presented for them and request transparency about the processes used to generate feedback [61]. If analytics are intended to serve learners, they should have opportunities to be involved in its implementation and have agency in its use [8, 31, 45]. Data from learning management systems may be easily accessible but it is important that this data are used ethically, responsibly, and in ways that are clear to learners both for research and classroom use (for further details see [87]). Learners can play a role in how analytics feedback is interpreted and have a say in the amount of guidance needed [82]. Providing guidance around how the analytics offered can support learning is a crucial piece that is often overlooked in the implementation design [45, 81].

Analytics should be adaptable, tailoring feedback to the needs of diverse learners. We caution learning analytics designers and those using analytics in their courses to take issues of accessibility into consideration, by designing or selecting visual representations of data that follow the guidelines for Accessible Rich Internet Applications [85]. For example, consider providing screen reader accessible descriptions of the analytic visualizations and using color combinations all viewers can distinguish [23]. Tools used to provide feedback should cater to the varying needs of learners. This includes providing sufficient time to make responses or removing anxiety triggers which can reduce barriers to accessing and processing analytic feedback [39]. The design of feedback and data used to generate it should reflect the diversity in the population it serves.

4 Future Directions and Conclusion

An avenue for research could focus on incorporating learner intentionality into learning analytics. This would aid in contextualizing learner data and offer guidance on how and when to generate feedback. A promising approach to capture learner intentions is through learner self-set goals. Providing learners an opportunity to set goals and monitor goal progress is a productive time management strategy that enables tracing learners' plans and intended actions to achieve a goal (e.g., [32, 42, 47]). In an online learning environment, learners' actions can be traced and compared to their goals (e.g., [32]). These data highlight learning activities a learner carried out, but also uncover planned actions that were not acted on, data that would be overlooked when focusing merely on trace data.

Goals capture learners' intentions and plans in the context of the learning activity. Data harvested across different sources can reveal how learners follow through with their goals. For example, creating critique notes or applying *strength* and *weakness* tags to academic articles for the goal of writing an article critique indicates following through with that goal. Spending time on social media would be an indication for procrastination on the article critique. However, before intervening, additional data about emotional engagement is needed to indicate if a learner is delaying a task to momentarily reduce fear of failure or is confused on how to critique the information provided. Multimodal data, sentiment analyses of discussion boards or tags learners apply to information sources can indicate a learner experiences fear in relation to the delayed assignment or confusion around the topic (e.g., [18, 22, 58]). These data are most informative when focusing on sequences of engagement, reflecting how individual learners learn.

To provide context for learning goals, data collection methods can be refined to more fully capture features of the environment where learning occurs [80]. Actions undertaken by the learner are caried out in the setting of the learning environment. Data collected is thus a reflection of how the learner interacted with the instructions and deadline provided, affordances of the system, and personal goals for learning. Actions should be indexed in relation to important qualities of the environment.

Pedagogically, considerations should be made for the learners' beliefs and understanding about the nature, origin, and limitations of human knowledge, such as knowledge acquisition, problem-based learning, construction of conceptual understanding, collaborative learning environments etc. which frame the learning activities. The design of activities can be catalogued as well as the tools available, how they are design and used. These are important considerations when making inferences about learner's intentions and designing feedback to guide learner engagement [70, 82]. If the analytic feedback interventions provided are successful, recording these factors will aid in making meaningful generalizations of the findings to other contexts [80].

In addition to contextualizing data collection and interpretation, there is potential for goal setting to serve as a venue for learners to make choices and personalize analytics feedback. Drawing on learners' goals is a promising way to provide meaningful feedback and recommendations. Framing analytics around goals affords personalization to individual learners' learning plans and intentions. Analytics feedback can reflect goal progress and provide recommendations of strategies available in the learning environment that are conducive to learning and goal achievement [32]. Future research can leverage learners' goals as a waypoint for designing analytics feedback. The format and timing of analytics feedback could be linked to learner goals and planning. Learners could be actively involved in this process. A learning analytics interface could be generated to allow learners to choose what data are used, how it is visualized, and when analytics are provided. This affords personalization of feedback to learner goals and facilitates autonomy. Learners are agents over their learning. Putting learners in charge of generating their own feedback promotes monitoring and reflection about their learning (e.g., [44]) and ensures meaningful feedback. Learning environments that facilitate autonomy are also associated with motivational benefits for learners [10, 77].

Incorporating learner goals provides educators greater insight into how learners are approaching tasks that could not be inferred from data alone. For example, trace data alone would not indicate whether a learner is aware of a strategy needed to accomplish a task but did not execute it well (indicating a potential utilization deficiency, [72, 73]) or the learner failed to recognize a strategy they could have used (indicating a potential production deficiency, [72, 73]). These are important considerations that can make recommendations provided through learning analytics more actionable and meaningful.

Although the field of learning analytics has made substantial contributions to harvesting and leveraging data about behavioral, cognitive, and emotional engagement of learners for analytics, there is yet more to learn about why and how learners make choices about learning. This specificity is required for operationalizations of complex, multi-faceted constructs such as procrastination. Future research is needed to refine data collection methods and generate operational definitions for learners' emotional engagement. This exploration could result in more systematic and impactful learning analytics.

References

1. Akram, A., Fu, C., Li, Y., Javed, M.Y., Lin, R., Jiang, Y., Tang, Y.: Predicting students' academic procrastination in blended learning course using homework submission data. IEEE Access **7**, 102487–102498 (2019). https://doi.org/10.1109/ACCESS.2019.2930867
2. Amon, M.J., Vrzakova, H., D'Mello, S.K.: Beyond dyadic coordination: multimodal behavioral irregularity in triads predicts facets of collaborative problem solving. Cogn. Sci. **43**(10), 1–22 (2019). https://doi.org/10.1111/cogs.12787
3. Baker, R.S.: Stupid tutoring systems, intelligent humans. Int. J. Artif. Intell. Educ. **26**(2), 600–614 (2016). https://doi.org/10.1007/s40593-016-0105-0
4. Baker, S., Inventado, P.S.: Educational data mining and learning analytics: Potentials and possibilities for online education. In: Veletsianos, G. (ed.) Emergence and Innovation in Digital Learning, pp. 83–98. Athabasca University Press (2016). https://doi.org/10.15215/aupress/978 1771991490.01
5. Blikstein, P., Worsley, M.: Multimodal learning analytics and education data mining: using computational technologies to measure complex learning tasks. J. Learn. Anal. **3**(2), 220–238 (2016). https://doi.org/10.18608/jla.2016.32.11
6. Bodily, R., Verbert, K.: Review of research on student-facing learning analytics dashboards and educational recommender systems. IEEE Trans. Learn. Technol. **10**(4), 405–418 (2017). https://doi.org/10.1109/TLT.2017.2740172
7. Brown, E., Glover, C.: Evaluating written feedback. In: Bryan, C., Clegg, K. (eds.) Innovative Assessment in Higher Education, pp. 81–91. Routledge (2006)
8. Buckingham Shum, S., Ferguson, R., Martinez-Maldonado, R.: Human-centered learning analytics. J. Learn. Anal. **6**(2), 1–9 (2019). https://doi.org/10.18608/jla.2019.62.1
9. Claessens, B.J., Van Eerde, W., Rutte, C.G., Roe, R.A.: A review of the time management literature. Pers. Rev. **36**(2), 255–276 (2007). https://doi.org/10.1108/00483480710726136
10. Cleary, T.J.: The Self-Regulated Learning Guide: Teaching Students to Think in the Language of Strategies. Routledge (2018)
11. Clow, D.: An overview of learning analytics. Teach. High. Educ. **18**(6), 683–695 (2013). https://doi.org/10.1080/13562517.2013.827653
12. Dawson, P., Henderson, M., Mahoney, P., Phillips, M., Ryan, T., Boud, D., Molloy, E.: What makes for effective feedback: staff and student perspectives. Assess. Eval. High. Educ. **44**(1), 25–36 (2019). https://doi.org/10.1080/02602938.2018.1467877
13. D'Mello, S.: A selective meta-analysis on the relative incidence of discrete affective states during learning with technology. J. Educ. Psychol. **105**(4), 1082–1099 (2013). https://doi.org/10.1037/a0032674
14. Duval, E., Verbert, K.: Learning analytics. ELEED: E-Learn. Educ. **8**(1), 3–7 (2012)
15. Ez-Zaouia, M., Lavoué, E.: EMODA: a tutor oriented multimodal and contextual emotional dashboard. In: Proceedings of the Seventh International Learning Analytics & Knowledge Conference, pp. 429–438 (2017)
16. Fairclough, S.H., Venables, L., Tattersall, A.: The influence of task demand and learning on the psychophysiological response. Int. J. Psychophysiol. **56**(2), 171–184 (2005). https://doi.org/10.1016/j.ijpsycho.2004.11.003
17. Ferguson, R.: Learning analytics: drivers, developments and challenges. Int. J. Technol. Enhanced Learn. **4**(5/6), 304–317 (2012). https://doi.org/10.1504/IJTEL.2012.051816
18. Gelmez, K., Bagli, H.: Tracing design students' affective journeys through reflective writing. Int. J. Technol. Des. Educ. **28**(4), 1061–1081 (2018). https://doi.org/10.1007/s10798-017-9424-1
19. Grunschel, C., Patrzek, J., Fries, S.: Exploring reasons and consequences of academic procrastination: an interview study. Eur. J. Psychol. Educ. **28**(3), 841–861 (2013). https://doi.org/10.1007/s10212-012-0143-4
20. Henderson, M., Ajjawi, R., Boud, D., Molloy, E.: Identifying feedback that has impact. In: M. Henderson, R. Ajjawi, D. Boud, E. Molloy (eds.) The Impact of Feedback in Higher Education, pp 15–34 (2017). Springer. https://doi.org/10.1007/978-3-030-25112-8163_2

21. Hooshyar, D., Pedaste, M., Yang, Y.: Mining educational data to predict students' performance through procrastination behavior. Entropy **22**(1), 12–36 (2020). https://doi.org/10.3390/e22 010012
22. Hu, A., Flaxman, S.: Multimodal sentiment analysis to explore the structure of emotions. In: Proceedings of the 24th ACM SIGKDD International Conference on Knowledge Discovery & Data Mining, pp. 350–358 (2018). https://doi.org/10.1145/3219819.3219853
23. Ichihara, Y.G., Okabe, M., Iga, K., Tanaka, Y., Musha, K., Ito, K.: Color universal design: the selection of four easily distinguishable colors for all color vision types. In: Proceedings of SPIE, vol. 6807 (2008). https://doi.org/10.1117/12.765420
24. Ishii, R., Otsuka, K., Kumano, S., Higashinaka, R., Tomita, J.: Analyzing gaze behavior and dialogue act during turn-taking for estimating empathy skill level. In: Proceedings of the 20th ACM International Conference on Multimodal Interaction, pp. 31–39 (2018). https://doi.org/ 10.1145/3242969.3242978
25. Järvelä, S., Malmberg, J., Haataja, E., Sobocinski, M., Kirschner, P.A.: What multimodal data can tell us about the students' regulation of their learning process. Learn. Instr. **72**, 101203 (2019). https://doi.org/10.1016/j.learninstruc.2019.04.004
26. Jivet, I., Scheffel, M., Schmitz, M., Robbers, S., Specht, M., Drachsler, H.: From students with love: an empirical study on learner goals, self-regulated learning and sense-making of learning analytics in higher education. Internet Higher Educ. **47**, 100758 (2020). https://doi. org/10.1016/j.iheduc.2020.100758
27. Joksimović, S., Kovanović, V., Dawson, S.: The journey of learning analytics. HERDSA Rev. Higher Educ. **6**, 27–63 (2019)
28. Kahn, P., Everington, L., Kelm, K., Reid, I., Watkins, F.: Understanding student engagement in online learning environments: The role of reflexivity. Educ. Technol. Res. Dev. **65**, 203–218 (2017). https://doi.org/10.1007/s11423-016-9484-z
29. Kahu, E., Nelson, K., Picton, C.: Student interest as a key driver of engagement for first year students. Stud. Success **8**(2), 55–66 (2017). https://doi.org/10.5204/ssj.v8i2.379
30. Klingsieck, K.B., Fries, S., Horz, C., Hofer, M.: Procrastination in a distance university setting. Distance Educ. **33**(3), 295–310 (2012). https://doi.org/10.1080/01587919.2012.723165
31. Knight, S., et al.: AcaWriter: a learning analytics tool for formative feedback on academic writing. J. Writ. Res. **12**(1), 299–344 (2020). https://doi.org/10.17239/jowr-2020.12.01.06
32. Konert, J., Bohr, C., Bellhäuser, H., Rensing, C.: PeerLA-assistant for individual learning goals and self-regulation competency improvement in online learning scenarios. In: IEEE 16th International Conference on Advanced Learning Technologies, pp. 52–56. IEEE (2016)
33. Kotsopoulos, D.: Developing an undergraduate business course using open educational resources. Can. J. Scholarsh. Teach. Learn. **13**(1) (2022). https://doi.org/10.5206/cjsotlrcacea. 2022.1.10992
34. Kovanovic, V., Gašević, D., Dawson, S., Joksimovic, S., Baker, R.: Does time-on-task estimation matter? implications on validity of learning analytics findings. J. Learn. Anal. **2**(3), 81–110 (2016). https://doi.org/10.18608/jla.2015.23.6
35. LAK.: 1st International Conference on Learning Analytics and Knowledge, Banff, AB, Canada (2011)
36. Larmuseau, C., Vanneste, P., Desmet, P., Depaepe, F.: Multichannel data for understanding cognitive affordances during complex problem solving. In: Proceedings of the 9th International Conference on Learning Analytics & Knowledge, pp. 61–70 (2019). https://doi.org/10.1145/ 3303772.3303778
37. Laney, D.: 3D data management: controlling data volume, velocity and variety. META Group Res. Note **6**, 70 (2001)
38. Lindblom-Ylänne, S.A., Saariaho-Räsänen, E.J., Inkinen, M.S., Haarala-Muhonen, A.E., Hailikari, T.K.: Academic procrastinators, strategic delayers and something betwixt and between: an interview study. Front. Learn. Res. **3**(2), 47–62 (2015). https://doi.org/10.14786/ flr.v3i2.154
39. Lister, K., Coughlan, T., Iniesto, F., Freear, N., Devine, P.: Accessible conversational user interfaces: considerations for design. In: Proceedings of the 17th International Web for All Conference, pp. 1–11 (2020)

40. Lockyer, L., Heathcote, E., Dawson, S.: Informing pedagogical action: aligning learning analytics with learning design. Am. Behav. Sci. **57**(10), 1439–1459 (2013). https://doi.org/10.1177/0002764213479367
41. Macfadyen, L.P., Lockyer, L., Rienties, B.: Learning design and learning analytics: Snapshot. J. Learn. Anal. **7**(3), 6–12 (2020). https://doi.org/10.18608/jla.2020.73
42. Mai, N.N., Takahashi, Y., Oo, M.M.: Testing the effectiveness of transfer interventions using Solomon four-group designs. Educ. Sci. **10**(4), 92–106 (2020). https://doi.org/10.3390/educsci10040092
43. Marzouk, Z., Rakovic, M., Liaqat, A., Vytasek, J., Samadi, D., Stewart-Alonso, J., Nesbit, J.C.: What if learning analytics were based on learning science? Australas. J. Educ. Technol. **32**(6), 1–18 (2016). https://doi.org/10.14742/ajet.3058
44. McCardle, L., Webster, E.A., Haffey, A., Hadwin, A.F.: Examining students' self-set goals for self-regulated learning: goal properties and patterns. Stud. High. Educ. **42**(11), 2153–2169 (2017). https://doi.org/10.1080/03075079.2015.1135117
45. Ochoa, X., Wise, A.F.: Supporting the shift to digital with learner-centered learning analytics. Educ. Tech. Res. Dev. **69**, 357–361 (2021). https://doi.org/10.1007/s11423-020-09882-2
46. Ochoa, X., Worsley, M.: Augmenting learning analytics with multimodal sensory data. J. Learn. Anal. **3**(2), 213–219 (2016). https://doi.org/10.18608/jla.2016.32.10
47. Papamitsiou, Z., Economides, A.A.: Exploring autonomous learning capacity from a self-regulated learning perspective using learning analytics. Br. J. Edu. Technol. **50**(6), 3138–3155 (2019). https://doi.org/10.1111/bjet.12747
48. Patzak, A.: Measuring and understanding self-handicapping in education. Unpublished doctoral dissertation. Simon Fraser University (2020)
49. Rebetez, M.M.L., Rochat, L., Van der Linden, M.: Cognitive, emotional, and motivational factors related to procrastination: a cluster analytic approach. Pers. Individ. Differ. **76**, 1–6 (2015). https://doi.org/10.1016/j.paid.2014.11.044
50. Rienties, B., Cross, S., Zdrahal, Z.: Implementing a learning analytics intervention and evaluation framework: what works? In: Motidyang, B., Butson, R. (eds.) Big Data and Learning Analytics in Higher Education, pp. 147–166. Springer, Cham (2017). https://doi.org/10.1007/978-3-319-06520-5_10
51. Rienties, B., Rivers, B.A.: Measuring and understanding learner emotions: evidence and prospects. Learn. Anal. Rev. **1**(1), 1–27 (2014)
52. Rogaten, J., Rienties, B., Sharpe, R., Cross, S., Whitelock, D., Lygo-Baker, S., Littlejohn, A.: Reviewing affective, behavioural and cognitive learning gains in higher education. Assess. Eval. High. Educ. **44**(3), 321–337 (2019). https://doi.org/10.1080/02602938.2018.1504277
53. Ruiz, S., Charleer, S., Urretavizcaya, M., Klerkx, J., Fernández-Castro, I., Duval, E.: Supporting learning by considering emotions: tracking and visualization a case study. In: Proceedings of the Sixth International Conference on Learning Analytics & Knowledge, pp. 254–263 (2016)
54. Samuelsen, J., Chen, W., Wasson, B.: Integrating multiple data sources for learning analytics—review of literature. Res. Pract. Technol. Enhanc. Learn. **14**(1), 1–20 (2019). https://doi.org/10.1186/s41039-019-0105-4
55. Santelli, B., Robertson, S.N., Larson, E.K., Humphrey, S.: Procrastination and delayed assignment submissions: student and faculty perceptions of late point policy and grace within an online learning environment. Online Learn. **24**(3), 35–49 (2020)
56. Sarmiento, J.P., Wise, A.F.: Participatory and co-design of learning analytics: an initial review of the literature. In: LAK22: 12th International Learning Analytics and Knowledge Conference, pp. 535–541 (2022)
57. Shankar, S.K., Prieto, L.P., Rodriguez-Triana, M.J., Ruiz-Calleja, A.: A review of multimodal learning analytics architectures. In: 2018 IEEE 18th International Conference on Advanced Learning Technologies (ICALT), vol. 00, pp. 212–214 (2018). https://doi.org/10.1109/ICALT.2018.00057
58. Soleymani, M., Garcia, D., Jou, B., Schuller, B., Chang, S.F., Pantic, M.: A survey of multimodal sentiment analysis. Image Vis. Comput. **65**, 3–14 (2017). https://doi.org/10.1016/j.imavis.2017.08.003

59. Siemens, G.: Learning analytics: the emergence of a discipline. Am. Behav. Sci. **57**(10), 1380–1400 (2013). https://doi.org/10.1177/0002764213498851
60. Simpson, W.K., Pychyl, T.A.: In search of the arousal procrastinator: Investigating the relation between procrastination, arousal-based personality traits and beliefs about procrastination motivations. Personal. Individ. Differ. **47**(8), 906–911 (2009). https://doi.org/10.1016/j.paid.2009.07.013
61. Slade, S., Prinsloo, P.: Student perspectives on the use of their data: Between intrusion, surveillance, and care. In: Challenges for research into open & distance learning: doing things better—doing better things, pp. 291–300. Eur. Distance E-Learn. Netw., Oxford (2014)
62. SOLAR.: What is Learning Analytics? Retrieved July 2021 from (2021). https://www.solaresearch.org/about/what-is-learning-analytics/
63. Steel, P.: The nature of procrastination: a meta-analytic and theoretical review of quintessential self-regulatory failure. Psychol. Bull. **133**(1), 65–94 (2007). https://doi.org/10.1037/0033-2909.133.1.65
64. Steel, P., König, C.J.: Integrating theories of motivation. Acad. Manag. Rev. **31**(4), 889–913 (2006). https://doi.org/10.5465/amr.2006.22527462
65. Stevens, S.S.: The operational definition of psychological concepts. Psychol. Rev. **42**(6), 517–527 (1935). https://doi.org/10.1037/h0056973
66. Stewart, A.E., Keirn, Z.A., D'Mello, S.K.: Multimodal modeling of coordination and coregulation patterns in speech rate during triadic collaborative problem solving. In: Proceedings of the 20th ACM International Conference on Multimodal Interaction, pp. 21–30 (2018). https://doi.org/10.1145/3242969.3242989
67. Tuominen, H., Niemivirta, M., Lonka, K., Salmela-Aro, K.: Motivation across a transition: changes in achievement goal orientations and academic well-being from elementary to secondary school. Learn. Individ. Differ. **79**, 101854 (2020). https://doi.org/10.1016/j.lindif.2020.101854
68. van Leeuwen, A.: Learning analytics to support teachers during synchronous CSCL: balancing between overview and overload. J. Learn. Anal. **2**(2), 138–162 (2015). https://doi.org/10.18608/jla.2015.22.11
69. Vrzakova, H., Amon, M.J., Stewart, A., Duran, N.D., D'Mello, S.K.: Focused or stuck together: multimodal patterns reveal triads' performance in collaborative problem solving. In: Proceedings of the Tenth International Conference on Learning Analytics & Knowledge, pp. 295–304 (2020). https://doi.org/10.1145/3375462.3375467
70. Vytasek, J.M., Patzak, A., Winne, P.H.: Analytics for student engagement. In: Virvou, M., Alepis, E., Tsihrintzis, G.A., Jain, L.C. (eds.) Machine Learning Paradigms, pp. 23–48. Springer, New York, NY (2020). https://doi.org/10.1007/978-3-030-13743-4_3
71. Wang, M.T., Binning, K.R., Del Toro, J., Qin, X., Zepeda, C.D.: Skill, thrill, and will: the role of metacognition, interest, and self-control in predicting student engagement in mathematics learning over time. Child Dev. **92**(4), 1369–1387 (2021). https://doi.org/10.1111/cdev.13531
72. Winne, P.H.: Experimenting to bootstrap self-regulated learning. J. Educ. Psychol. **89**(3), 397–410 (1997). https://doi.org/10.1037/0022-0663.89.3.397
73. Winne, P.H.: Bootstrapping learner's self-regulated learning. Psychol. Test Assess. Model. **52**(4), 472–490 (2010)
74. Winne, P.H.: Construct and consequential validity for learning analytics based on trace data. Comput. Hum. Behav. **112**, 106457 (2020). https://doi.org/10.1016/j.chb.2020.106457
75. Winne, P.H.: Commentary: a proposed remedy for grievances about self-report methodologies. Front. Learn. Res. **8**(3), 164–173 (2020b). https://doi.org/10.14786/flr.v8i3.625
76. Winne, P.H., Teng, K., Chang, D., Lin, M.P.C., Marzouk, Z., Nesbit, J.C., Patzak, A., Rakovic, M., Samadi, D., Vytasek, J.: nStudy: Software for learning analytics about processes for self-regulated learning. J. Learn. Anal. **6**(2), 95–106 (2019). https://doi.org/10.18608/jla.2019.62.7
77. Winne, P.H., Marzouk, Z.: Learning strategies and self-regulated learning (Ch. 27). In: Dunlosky, J., Rawson, K.A. (eds) The Cambridge Handbook of Cognition and Education, pp 696–715. Cambridge University Press (2019)

78. Winstone, N., Carless, D.: Designing Effective Feedback Processes in Higher Education: A Learning-Focused Approach. Routledge (2019)
79. Wise, A.F.: Designing pedagogical interventions to support student use of learning analytics. In: Proceedings of the Fourth International Conference on Learning Analytics and Knowledge, pp. 203–211 (2014)
80. Wise, A.F.: Learning analytics: using data-informed decision-making to improve teaching and learning. In: Adesope, O.O., Rud, A.G. (eds.) Contemporary technologies in education, pp. 119–143. Palgrave Macmillan, Cham (2019). https://doi.org/10.1007/978-3-319-896 80-9_7
81. Wise, A.F., Shaffer, D.W.: Why theory matters more than ever in the age of big data. J. Learn. Anal. 2(2), 5–13 (2015). https://doi.org/10.18608/jla.2015.22.2
82. Wise, A.F., Vytasek, J.M.: Learning analytics implementation design. In: Lang, C., Siemens, G., Wise, A., Gašević, D. (eds.) Handbook of Learning Analytics and Educational Data Mining, pp. 151–160. Society for Learning Analytics Research (2017). https://doi.org/10.18608/hla17
83. Wise, A.F., Vytasek, J.M., Hausknecht, S., Zhao, Y.: Developing learning analytics design knowledge in the "middle space": the student tuning model and align design framework for learning analytics use. Online Learn. J. 20(2):155–182 (2016). https://doi.org/10.24059/OLJ. V20I2.783
84. Worsley, M.: (Dis)engagement matters: Identifying efficacious learning practices with multimodal learning analytics. In: Proceedings of the 8th International Conference on Learning Analytics and Knowledge, pp. 365–369 (2018)
85. W3C.: Web Content Accessibility Guidelines (WCAG) 2.1. Accessed July 2021 from http://www.w3.org/TR/WCAG/ (2018)
86. Tempelaar, D., Nguyen, Q., Rienties, B.: Learning feedback based on dispositional learning analytics. In: Virvou, M., Alepis, E., Tsihrintzis, G.A., Jain, L.C. (eds) Machine learning paradigms: Advances in learning analytics. Intelligent systems reference library, vol. 158, pp. 69–89. Springer, Cham (2019). https://doi.org/10.1007/978-3-030-13743-4_5
87. Tzimas, D., Demetriadis, S.: Ethical issues in learning analytics: a review of the field. Educ. Tech. Res. Dev. 69(2), 1101–1133 (2021). https://doi.org/10.1007/s11423-021-09977-4

Improving Data Literacy in Management Education Through Experiential Learning: A Demonstration Using Tableau Software

Aria Teimourzadeh and Samaneh Kakavand

Abstract Problem solving skills using advanced analytics techniques and tools have become essential in today's era of ever increasing data. This necessity in the field of management pushes those students who do not have a data science background to acquire skills related to the use of new tools and methods that continue to evolve. Previous research highlights many technical topics in data science and analytics that are limited to conceptual discussions in various management courses. The main instructional challenges in management are related to the cost of different software, students' low data literacy and also their lack of technical skills. In the context of management, the data analytics workflow starts with the definition of a business problem and it ends with the creation of relevant interactive reports and dashboards that containing analysis results. This chapter provides a guideline on the integration of data analytics in management education, framed in the context of the stages of such a workflow. This chapter illustrates examples to management educators that allow them to demonstrate different stages of data analytics workflow using Tableau software.

Keywords Analytical skills · Data literacy · Data analysis · Data visualization · Tableau

1 Introduction

The technological revolution has transformed the way various managerial decisions are made [10]. Given that many managerial roles have been shifting towards evidence-based decisions and practices, scholars have highlighted the necessity of integrating

A. Teimourzadeh (✉)
Huron at Western, The University of Western Ontario, London, ON, Canada
e-mail: ateimour@uwo.ca

S. Kakavand
Wilfrid Laurier University, Waterloo, ON, Canada
e-mail: skakavand@wlu.ca

© The Author(s), under exclusive license to Springer Nature Switzerland AG 2023   83
D. G. Woolford et al. (eds.), *Applied Data Science*, Studies in Big Data 125,
https://doi.org/10.1007/978-3-031-29937-7_7

data analytics in management curriculum to better adapt the graduates' skills to the needs of business organizations [2].

In today's world, a large amount of data is being generated every second in each and every sector of the economy and society. This has pushed the scientific community and computer engineers to focus on the creation of advanced software and techniques that allow the users to collect, store, process, and analyze large and complex data sets. The need to extract structure and patterns requires the availability of high-quality data and use of advanced software and techniques. In business-oriented environments, a high number of managerial and strategic decisions are heavily based on insights from data analyses related to core business processes and consumer behavior.

In academic institutions, enabling students to acquire evidence-based decision-making skills using automated tools adapts the students' knowledge to the job market needs [4]. It is noteworthy that innovative experiential learning methods will facilitate knowledge acquisition about real-life applications of business concepts. Consequently, management educators tend to incorporate experiential learning in the classroom, so students can work with real business data sets. In some business schools, lack of specific teaching guidelines related to data analytics, proper resources, cost of software [6] and students' lack of technical skills continue to be real challenges for both faculty and students. Given the obstacles for the integration of data science and analytics tools, the discovery of available software that can contribute to learning objectives of big data analytics in management programs is crucial. The main learning objectives in a typical data-driven management course include business problem identification, data collection, data processing, data analysis, data visualization and data translation.

This chapter provides both theoretical as well as practical hands-on experience using advanced data analysis and data visualization software. In the context of management education, this includes instructions on how to improve data literacy using two user-friendly software applications known as Tableau Prep Builder and Tableau Desktop.

The content of this chapter will help educators to familiarize themselves with relevant data collection, data transformation, data analysis, data visualization and reporting methods. Their efforts in this experiential learning method will enable users (i.e., students) to acquire analytical skills through user-friendly programs.

2   Data Literacy

Scholars have examined the concept of data literacy, but its definition varies depending on the field studied. According to [7], data literacy is the ability to understand and use data effectively for decision making. This definition focuses on using data and not on the necessary technical aspects. Among the definitions related to analytical skills, some scholars focus primarily on mathematical, computer and statistical skills [15]. It seems that transversal skills based on knowledge such as the epistemology of big data, the legal framework, project management [8], governance, quality, and data ethics [9] are essential to mastering data literacy.

According to Calzada-[12], the concept of data literacy is the ability of individuals to access, interpret, evaluate, manage, manipulate and use data ethically. In this sense, the challenge of data literacy is the development of a data-oriented culture and the ability to ethically and critically evaluate data [1].

In the extant research, [15] have analyzed different definitions of data literacy as well as statistical literacy to find similarities among those definitions. Based on the result of their analysis, data literacy refers to:

> *the ability to ask and answer real-world questions from large and small data sets through an inquiry process, with consideration of ethical use of data. It is based on core practical and creative skills, with the ability to extend knowledge of specialist data handling skills according to goal. These include the abilities to select, clean, analyze, visualize, critique, and interpret data, as well as to communicate stories from data and to use data as part of a design process.* (p. 23)

Among the attempts to define data literacy, statistical literacy has often been underestimated, suggesting that the quantity of data could be enough to provide answers to the various questions by focusing on the exploration and discovery of relationships [1]. However, quantity may mislead the users if they don't have all the necessary skills to process and analyze data as well as interpret the results.

According to a statistical literacy framework which was developed by [14], an analyst should have three types of abilities. The first one is the technical ability which is to understand statistical terminology. The second is social ability which is the ability to understand statistics in a boarder societal context. The third ability is related to critical thinking which is the most challenging skill. It is worth noting that critical thinking, evaluation and interpretation of data, is the key ability to improve a data-driven decision making.

Next, we give an overview about the concepts of big data and analytics in order to clarify the steps for becoming data literate through the support of statistical techniques in a management education context.

3 What is Big Data?

Big data has become a buzzword with different meanings to different users. Many business firms have recognized big data as an important source of information enabling further value-creation in operational, managerial and strategic decisions. Big data refers to high volume data collected by various sources such as electronic devices, sensors, satellites, social media feeds, and GPS signals [3]. The incapability of traditional software and lack of expertise to collect, transform, store, analyze and visualize big data remain challenges for many business firms. Hence, the exploration of this phenomenon requires the availability of advanced software as well as individuals who have data analysis and data translation skills.

Table 1 Characteristics of big data (5Vs) and examples

Characteristic	Example
Volume	This refers to data size. For instance, Alibaba's data warehouse contains more than 10 petabytes of transactional data
Velocity	This refers to the speed at which the data are being generated and has to be processed for analysis. For example, 4 million rows of financial data in a bank are being processed and stored in less than 5 hours
Variety	This includes different data structures. For example, Amazon has the ability to process and analyze customers' data which exist in different formats such as product images, videos and audio files uploaded by the customers
Veracity	This refers to the quality of data. For instance, Walmart improves its out-of-date customer database and replaces the old data using fresh and up-to-date data to increase the accuracy of decisions
Value	The higher the veracity, the more value will be generated. For instance, a firm collects relevant data from reliable data sources to improve online shopping experiences, which can increase sales

Oracle [11] has identified three main characteristics to better describe big data. These characteristics include volume, velocity, and variety. Volume is a term to describe the amount of data which continues to increase. Velocity refers to the speed at which the data should be transformed and prepared for analysis. Variety refers to different structures of data such as text, images, videos etc. Other characteristics such as veracity and value have been added to describe the quality, truthfulness of data and the value it generates for decision making. Table 1. illustrates examples for each characteristic of big data.

4 Data Analytics

In every analytics project, business firms must have a clear methodology to transform big data into valuable information assets. Sometimes this involves methods for working with big data in real time. Sometimes this involves the analysis of other data that is large and complex, but not strictly big data, per se. The term "analytics" is commonly used to refer to the methods for working with, analyzing, and modelling such data. However, this term does not have a standard definition. In general, it refers to the management of the data workflow using scientific techniques and automated tools. As described in what follows, this data analytics workflow consists of several stages, encompassing aspects from business problem identification through to collecting, processing, analyzing and visualizing data, followed by interpreting results. All these tasks are carried out to prepare data for current and future analysis and the ultimate goal is to support decision making.

5   The Six Stages of the Data Analytics Workflow

5.1   Business Problem Identification

This stage has been recognized as one of the most complex, but largely ignored stages in the data analytics workflow. Every big data analytics project requires a well-defined business problem that presents a clear analysis goal. This would help decision-makers to identify the resources that will need to be utilized. For instance, business firms might be interested in knowing why customers visit the website but do not buy their products. Another example could be an increasing number of support calls from a specific region.

5.2   Data Collection

Data collection includes the process of collecting data sets from both internal and external sources of data. It is essential to identify the data sources based on the analytics problem which has been identified in the previous stage. The purpose is to ensure that the collected data are relevant to business problem which plays a critical role in improving the accuracy of decisions.

5.3   Data Processing

The main purpose of data processing is to prepare data for analysis by compiling, cleaning, organizing and filtering data in preparation for analysis and visualization. In any knowledge discovery process, the value of the knowledge extracted is related to the quality of the data used. A common problem affecting quality is the existence of noisy or incomplete data. For instance, a data set may contain a lot of irrelevant characters or misspelled words. This may affect the quality of analysis results. Additionally, there could be a lot of unlabeled data, which makes it difficult to identify patterns. Considering the inconsistencies, incompleteness and duplication of raw data, organizations cannot extract hidden patterns from data efficiently without this processing step.

Basically, data processing includes the application of methods and operations in order to clean, filter, transform and classify raw data into valuable and organized information assets for further analysis. Given the increasing volume of unstructured data, the use of advanced automated tools and machine learning algorithms can facilitate this process.

5.4 Data Analysis

Data analysis includes a collection of methods and techniques to examine a data set and to make inferences about data. In other words, it is an investigation process in which patterns, structure and relationships will be revealed. In the analytics process, basic analysis focuses on the measures of central tendency such as mean, median, and mode as well as the measures of dispersion such as standard deviation and variance. These are also known as descriptive statistics that summarize the characteristics of a data set. Advanced analysis techniques that are widely used include classification, regression, clustering, anomaly detection and association analysis. These methods are mainly related to inferential statistics that focus on making predictions and drawing conclusions about a sample data set.

5.5 Data Visualization

According to [5], data visualization is defined as the process of translating large data sets and metrics into charts, graphs and other visuals. The visual representation of data is about ways to depict data through a choice of physical forms. By combining statistics and design, the aim of data visualization is to communicate information effectively to the readers. Typically, transformed data are visualized in the form of a chart, infographic, diagram or map. In a quantitative analysis, the statistical results are often visualized in a table to understand the strength of the evidence from the sample.

5.6 Data Translation

Data translation refers to the process of generating knowledge from the collected information and determining the significance, conclusions, implications, and possibly the application of the findings. The ability to interpret the results of the analysis still remains a challenge for many management graduates, and statistical literacy skills play an important role in data translation. For instance, whether a p-value is less than a given threshold should not be used as a rule to declare a result "significant" and then make business or policy decisions; instead it should be interpreted as the strength of evidence as one part of a robust and reproducible data analysis (see [13]). The absence of interpretation skills might lead to misguidance in the decision-making process. Hence, this step requires statistical literacy to ensure more accurate evidence-based decisions.

   The data translation skills include reading, understanding, and summarizing a chart, graph or table. A data interpreter should have the ability to develop findings, conclusions and recommendations in order to answer the business problem that has been identified in stage 1. In other words, a data translator is in charge of looking for the title of the graph, the labels, trends, structures, correlations, or outliers. It is worth noting that data translation refers to the process of assigning meaning to the analysis findings. At this stage, the user should think about potential hypothesis on what could be the root cause or what consequences may exist.

Developing data translation skills requires understanding the underlying data such as where it comes from and how key values are calculated. There are several key aspects to better understand data translation. First, the user should be familiarized with the unfamiliar or new data before processing it mentally. This includes looking at the metrics, dimensions and source of data being displayed. Secondly, any visible pattern, omissions and outliers in the data should be highlighted. Thirdly, a data translator must understand how other elements such as potential biases and incorrect assumptions can shape the interpretation. For example, any bias, comparisons or causations must be evaluated to avoid any confusion with other elements such as correlations. Finally, curiosity is the final step is to determine if any additional questions should be formulated. In some situations, a data translator may have the ability to explore the data further. Therefore, new questions can be posed that require further analysis and time.

5.7 Lesson Plan

The following section of this chapter highlights the required software and practical instructions for a specific lesson plan in order to improve knowledge about each step of data analytics workflow.

5.7.1 Required Software

Tableau[1] is one of the fastest growing Business Intelligence and Analytics software developed by Pat Hanrahan, Christian Chabot, and Chris Stolte from Stanford University in 2003. Given its evolution and success, Tableau was acquired by Salesforce in an all-stock deal worth over 15 billion dollars. This software currently has different versions with various purposes and levels of functionality. For example, instructors can request a free license every year and students can benefit from a 1-year free license to activate two premium versions of Tableau. In this section, we introduce *Tableau Prep Builder* which is used to perform data cleaning and data processing tasks. Additionally, *Tableau Desktop* is used for data analysis and data visualization.

---

[1] https://www.tableau.com.

5.7.2   Assignment Design

The objective of this assignment is to introduce each step of the data analytics work-flow through Tableau software. This will develop students' data literacy, furthering the development of effective data translation skills, in the context of management and organizational studies.

In this particular assignment, students learn about the functionality of *Tableau Prep* software to import, clean and organize the data. Additionally, they will perform data analysis and visualization tasks using *Tableau Desktop* software. It is noteworthy that this assignment does not require any specific prerequisite related to computer science or artificial intelligence. However, basic knowledge of statistics such as frequency tables, measures of central tendency, measures of dispersion and graphical displays of data would be beneficial to facilitate the learning process in data analysis and data visualization parts of the analytics workflow.

5.7.3   Learning Objectives

This assignment is designed to encourage students to:

* Recognize the opportunities of data analytics for managerial and strategic decision making
* Use Tableau Prep Builder and Tableau Desktop software
* Identify entities and attributes in the initial exploration of a large data set
* Distinguish the difference between relevant categorical and numerical variables in data analytics projects
* Develop technical skills related to cleaning tasks in order to prepare large data for analysis
* Classify data analysis and data visualization options
* Develop data translation skills related to quantitative data analysis

5.7.4   Teaching Instructions

Suggested time and assignment structure

Table 2 illustrates the designed plan for this assignment, which is designed for three 2-h classes.

5.7.5   Detailed Practical Instructions

Business problem identification and data collection

In this section, we emphasize the importance of specific business questions that can be formulated by the instructor prior to the class. A step-by-step guide is designed to show how students can connect to data sets for data preparation. One of the main

Table 2 Allocation of time and suggested assignment structure

Conceptual discussion	
Time	Suggested questions
30 min.	**Introduction** Why evidence-based decisions are important in business-oriented environments? What is big data? What are the characteristics of big data (5Vs)?
30 min.	What is data analytics? What is the data analytics workflow?
15 min.	What are some risks and opportunities of data analytics for business firms?
15 min.	What are the goals of data analytics in different firm functions such as marketing and sales, finance, operations?
Practical demonstration by the instructor	
Time	Suggested questions
15 min.	What is the purpose of business problem identification and why it is important in every analytics project?
20 min.	What is data collection? How to collect data from different sources? What are different data formats? Where to download open data sets?
40 min.	How to clean and process data? Do we need all the variables? Should we create new variables? Introduce Tableau Prep Builder for data preparation
60 min	How to analyze and visualize data using Tableau Desktop? Topics discussed focus on descriptive statistics, cluster analysis, correlation and regression analysis
Exercises to be completed by students during the class time	
Time	Tasks
15 min.	Downloading the data set provided by the instructor, importing the data set into Tableau Prep Builder, identifying the categorical and numerical variables
30 min.	(1) Initial exploration of a data set and data cleaning using Tableau Prep Builder (2) Previewing the transformed data set in Tableau Desktop
90 min.	Answering questions provided by the instructor using data analysis and visualization techniques

advantages of Tableau Prep Builder and Tableau Desktop is that they can deal with high variety data. In other words, the software allows users to import and read various data formats such as Excel, CSV, PDF, JSON and other semi-structured or structured data set which are mostly textual by nature.

For instance, a business company works towards different pricing strategies and improving its sales. One of the strategies could be promotions or discounts that may improve sales of certain product categories. A business question that can be answered using data in this context would be "Is there any correlation in the relationship between the two variables *discount* and *sales* in different product categories?".

Another example would be to identify if all product segments are profitable in terms of sales. Therefore, a business question can be "Is there any correlation in the relationship between the two variables *profit* and *sales* in different product segments?". In this case, we can see the trend for different product segments and identify the ones which are not profitable. Additionally, this would guide the analyst to focus on the aforementioned variables for further analysis.

In this assignment, a sample sales data set in Excel format can be provided to ensure that all students work with the same data set. Tableau software has included a sample superstore data set that can be used for demonstration and learning purposes. Figure 1 illustrates the data importing process in Tableau Prep Builder.

Data processing

Considering the tables in our sample data set, the users should double click on the 'Orders' table and click on 'view and clean' icon which is located at the top of the software interface. At this stage, the instructor explains how to verify data types in a data set using Tableau Prep Builder. For instance, a 'price' attribute must allow decimal places. The 'order date' must be in the appropriate date format. Furthermore, the importance of data cleaning and data transformation should be highlighted. Although some data sets may contain noisy values, the instructor should demonstrate the functionalities of Tableau Prep Builder to facilitate the data cleaning and transformation process. For instance, if the user decides to separate certain characters or values in a data set, the first step would be to click on the 'clean' icon as shown in

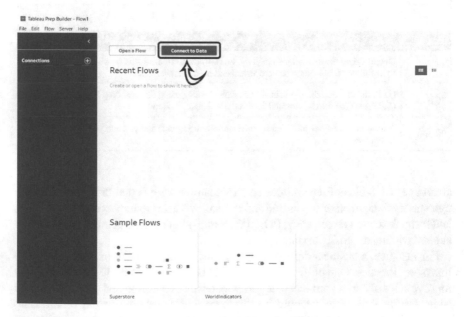

Fig. 1 Initial exploration of a data set using Tableau Prep Builder. © 2022 Tableau Software, LLC and its licensors

Fig. 2 Example of data transformation using Tableau Prep Builder. © 2022 Tableau Software, LLC and its licensors

Fig. 2. Additionally, the data can be converted to the desired format. For instance, a cell may contain the day, month and year. Tableau Prep Builder can extract the "month" value, if this value is the main focus for further analysis. Another option is to look for missing values in the filtering options. In certain situations, the user can avoid the data that is unnecessary using filter options.

At this specific stage, the data analyst should ask a few questions. (1) Do we need all the variables for analysis? (2) Should we convert certain categorical variables to numerical variables? (3) Are there any missing values? (4) Should we create any new variables? Fig. 2 illustrates an example of data cleaning and transformation.

Once the missing values have been identified and the cleaning process is completed, the transformed data can be previewed directly in Tableau Desktop. This is done by doing a right click on the 'clean' icon situated at the top and selecting 'Preview in Tableau Desktop'. Figure 3 illustrates how cleaned data can be previewed in Tableau Desktop.

Data analysis and data visualization

Prior to any data analysis and visualization exercise, a basic explanation related to the software interface is essential. Figure 4 illustrates the interface of Tableau Desktop and its basic functionalities.

1. This area shows the dimensions (mainly qualitative variables) of a table or combination of tables.
2. This area includes the measures or quantitative variables which are mainly used for calculation purposes. Tableau treats any field containing numeric (quantitative) information as a measure.

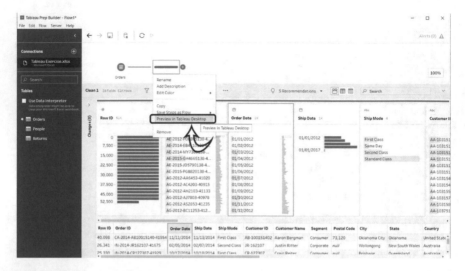

Fig. 3 Exporting clean data from Tableau Prep Builder to Tableau Desktop. © 2022 Tableau Software, LLC and its licensors

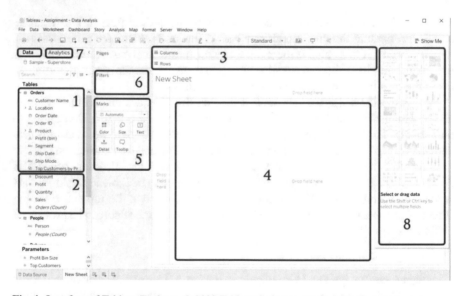

Fig. 4 Interface of Tableau Desktop. © 2022 Tableau Software, LLC and its licensors

3. Columns and Rows are the areas in which the user can drag and drop qualitative and quantitative variables for visualization purposes.
4. This area refers to the main visualization field in which tables, charts and maps will be displayed.

5. The user can drag and drop the variables in different fields such as colors, size and text in order to personalize the visualizations. For instance, the colors of tables and charts can be customized using this feature. Additionally, the user can drag and drop the variables into the text field to display the value they contain.
6. This area is used for filtering purposes. For example, users can filter the customer orders by date or by the amount of profit.
7. 'Data' tab allows users to observe all the detected tables and attributes of the imported data set. By clicking on the Analytics tab, the user will have the option to benefit from some prebuilt algorithms which can be used for descriptive statistics, forecasting, cluster analysis and other types of analyses.
8. This area allows users to select different types of chart, graph or table depending on the nature of analysis.

Figure 5 shows the example of an Excel file that has been imported to Tableau Desktop. This can be done if the user determines that the data set does not require prior cleaning and preparation in Tableau Prep Builder. Users should first import the data set and click on the 'Data Source' tab situated at the bottom left of the software to see the number of tables that are included in the data set. Additionally, they can choose to drag and drop a table or a combination of relational tables into the middle field and then click on a new worksheet to see all the variables.

Depending on the purpose of analysis, a list of relevant functions in Tableau Desktop can be demonstrated. Generally, simple and advanced calculations can be performed using a box in which formulas can be entered. This box is known as a 'calculated field'. The task can be accomplished by clicking on the 'Analysis' tab at the top of the screen. Then, the user should click on 'create calculated field'.

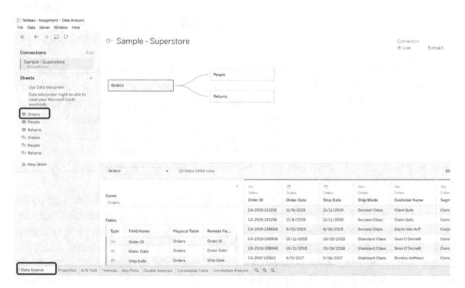

Fig. 5 Example of direct data set import process in Tableau Desktop. © 2022 Tableau Software, LLC and its licensors

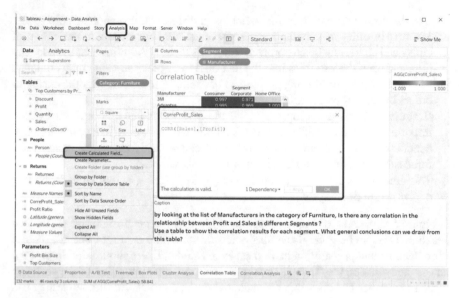

Fig. 6 Creation of calculated fields in Tableau Desktop. © 2022 Tableau Software, LLC and its licensors

As mentioned earlier, prior knowledge of introductory statistics can improve the students' learning for data analysis. Figure 6 demonstrates how calculated fields can be created for advanced calculations. In this particular example, the CORR function has been used to calculated the correlation between profit and sales in this data set.

Once a calculated field has been created, it will appear in the list of measures. Therefore, the user will be able to drag and drop this new measure into the visualization field. It should be noted that there are different ways of performing analysis in Tableau Desktop.

Figure 7 illustrates an example of correlation analysis using the 'Analytics' tab. This particular tab allows users to benefit from powerful features related to data modeling and descriptive statistics. For instance, they can drag and drop the 'segment' and 'sales' into columns and 'profit' into rows. To visualize individual values of quantitative variables, users should click on the 'analysis' tab and unselect 'aggregate measures'. The next step would be to drag and drop the 'category' into colors. The final step would be to drag and drop the 'Trend line' feature from the analytics tab into the visualization field.

It is important to note that there a variety visualization options depending on the nature of analysis. Tableau allows users to visualize data using different charts and graphs. Some examples include bar chart, line chart, pie chart, scatter plot, heat map, box and whiskers chart, highlight table, bubble chart, tree map, waterfall chart and histograms. Choosing a specific chart would depend on the nature of data and the type of analysis. We provide a few examples to clarify the choice of a chart or graph. For instance, a qualitative variable can be presented using a bar chart. The horizontal

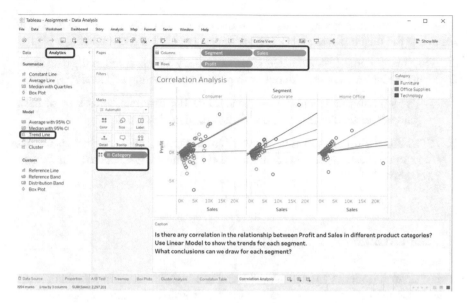

Fig. 7 Example of correlation analysis using Tableau Desktop. © 2022 Tableau Software, LLC and its licensors

axis of a bar chart highlights the classes corresponding to the variable of interest. The vertical axis of a bar chart shows the frequency of the observations in a data set. Bar charts are very effective charts for comparing magnitudes, and spotting highs and lows in the data.

A pie chart is another option to present qualitative data. A pie chart graphically describes the frequency of observations using a circle that is divided into slices. Each slice of the circle represents the relative frequency of observations as a percentage of the total number of observations. In other words, the size or angle of the slice represents that item or category's proportion to the whole. Pie charts are best at showing part-to-whole relationships when there are not too many slices.

A histogram is similar to a bar chart, but it is used when the analyst is dealing with quantitative data. Each bar of a histogram shows the distribution of quantitative data.

Our last example focuses on a scatter plot which is technically a collection of scattered points. Scatter plots require numeric values on both the vertical and horizontal axes. They highlight the relationship between two variables, and are great for visualizing clusters, showing possible correlations, and spotting outliers.

Analysis findings and discussions

As illustrated in Fig. 7, several worksheets can be created and different questions can be included in each worksheet. The purpose is to help students identify the variables which are relevant to the question being asked. Following the demonstration of each stage, the instructor can ask students to discuss the findings or recommendations related to the business questions. The answers should cover the following topics:

1. How many numerical and categorical variables were observed in the data set?
2. What variables were considered as important for cleaning and analysis?

   a. Which analysis technique has been used? For example, what type of basic and advanced charts have been chosen to visualize the data?

3. What are the findings of this analysis? In particular, what conclusions can we draw in the business context using the result of the analysis in different Tableau worksheets?

6 Conclusion

This chapter introduces an experiential learning method to develop data literacy in a data-driven management course. Given the importance of integrating data analytics in management education, it highlights the use of relevant tools and techniques to collect, process, analyze and visualize data in the context of marketing and sales activities.

Many business schools have been working to update the curriculum of management programs to improve student's employability and match their skills to business organizations' expectations. Considering that the cost of advanced software and automated tools remain one of the biggest challenges in academic institutions, this chapter proposes two powerful and free software programs that can be integrated in management courses to familiarize students with data analytics concepts. Additionally, the efforts in this pedagogical approach provide an innovative method of teaching data analytics in management discipline. The educators can enhance the learning process by providing step-by-step instructions as illustrated in this chapter. As a result, this process can help non-data science students acquire analytical skills through user-friendly programs.

References

1. Carlson, J., Fosmire, M., Miller, C.C., Nelson, M.S.: Determining data information literacy needs: a study of students and research faculty. Libr. Acad. **11**(2), 629–657 (2011). https://doi.org/10.1353/pla.2011.0022
2. Clayton, P.R., Clopton, J.: Business curriculum redesign: integrating data analytics. J. Educ. Bus. **94**(1), 57–63 (2019). https://doi.org/10.1080/08832323.2018.1502142
3. Davenport, T.H., Barth, P., Bean, R.: How big data is different. MIT Sloan Manag. Rev. **54**(1), 43–46 (2012)
4. Gillentine, A., Schulz, J.: Marketing the fantasy football league: utilization of simulation to enhance sport marketing concepts. J. Mark. Educ. **23**(3), 178–186 (2001). https://doi.org/10.1177/0273475301233003

5. IBM: What is data visualization? Retrieved 20 April 2022 (2021). https://www.ibm.com/ana
    lytics/data-visualization
6. Liu, X., Burns, A.C.: Designing a marketing analytics course for the digital age. Mark. Educ.
    Rev. **28**(1), 28–40 (2018). https://doi.org/10.1080/10528008.2017.1421049
7. Mandinach, E.B., Gummer, E.S.: A systemic view of implementing data literacy in educator
    preparation. Educ. Res. **42**(1), 30–37 (2013). https://doi.org/10.3102/0013189x12459803
8. Marr, B.: Big Data: Using SMART Big Data, Analytics and Metrics to Make Better Decisions
    and Improve Performance. John Wiley & Sons (2015)
9. Matthews, P.: Data literacy conceptions, community capabilities. J. Commun. Inform. **12**(3)
    (2016). https://doi.org/10.15353/joci.v12i3.3277
10. McAfee, A., Brynjolfsson, E., Davenport, T.H., Patil, D.J., Barton, D.: Big data: the
    management revolution. Harv. Bus. Rev. **90**(10), 60–68 (2012)
11. Oracle: What is Big Data? Retrieved 25 March 2022 (2012). https://www.oracle.com/ca-en/
    big-data/what-is-big-data/
12. Prado, J.C., Marzal, M.Á.: Incorporating data literacy into information literacy programs: core
    competencies and contents. Libri **63**(2), 123–134 (2013). https://doi.org/10.1515/libri-2013-
    0010
13. Wasserstein, R.L., Lazar, N.A.: The ASA statement on p-values: context, process, and purpose.
    Am. Stat. **70**(2), 129–133 (2016). https://doi.org/10.1080/00031305.2016.1154108
14. Watson, J.M.: Assessing statistical thinking using the media. In: Gal, I., Garfield, J. B. (eds.)
    The Assessment Challenge in Statistics Education, pp. 107–121 (1997)
15. Wolff, A., Moore, J., Zdráhal, Z., Hlosta, M., Kuzilek, J.: Data literacy for learning analytics.
    In: Proceedings of the Sixth International Conference on Learning Analytics & Knowledge,
    pp. 500–501 (2016)

Understanding Players and Play Through Game Analytics

Jonathan Tan, Mike Katchabaw, and Damir Slogar

Abstract Video games produce terabytes of data, and game studios are investing more and more in game analytics to parse that data, understand their players, and improve their games. However, an essential step in game analytics is to present the results to shareholders who may not be as familiar with the technical details of data science. In this chapter, we present a reusable methodology for game analytics and discuss how to translate the results from analyses using this methodology in an interpretable manner. To illustrate our method, we view game analytics through the lens of player clustering with a case study on the mobile game *My Singing Monsters*.

Keywords Game analytics · Cluster analysis · Player representation · Data visualization

1 Introduction

Video games can range from simple games, played with one finger on a mobile device, to more complex games with lengthy storylines and stunning graphics. In addition, the players who play these games are also just as unique. Take a giant open-world game such as Nintendo's *The Legend of Zelda: Breath of the Wild*, for example. Some might do the bare minimum to beat the final boss, while others look to acquire every collectible in the game. Understanding their players is vital for game companies to succeed, and, as a result, many engage in game analytics and adopt a data-driven approach to developing games [6, 7].

In this chapter, we discuss game analytics through the lens of work we did in [10], which performed a clustering analysis on the players of the game *My Singing Monsters*. Clustering involves the formation of clusters of data points grouped by their features. Each data point represents an individual player; thus, we formed clusters

J. Tan (✉) · M. Katchabaw
Department of Computer Science, Western University, London, ON, Canada
e-mail: jtan97@uwo.ca

D. Slogar
Big Blue Bubble, London, ON, Canada

© The Author(s), under exclusive license to Springer Nature Switzerland AG 2023
D. G. Woolford et al. (eds.), *Applied Data Science*, Studies in Big Data 125,
https://doi.org/10.1007/978-3-031-29937-7_8

based on player behavior. As a game analyst, identifying behaviors in the player base can lead to suggestions for game producers on modifying or creating new content to keep players happy.

However, while clustering may be a valuable exercise, successfully executing it is non-trivial given the sheer amount of data and complexity surrounding a video game and its players. Furthermore, a game studio comprises individuals from many different backgrounds, and translating the results of such analyses into an understandable and actionable form for non-technical parties is not easy either. Even among other data analysts, effectively communicating results is essential.

Thus, a guide through the process is needed. We hope this chapter can serve as one such guide, providing some insight from an analyst's point of view. The following section introduces the key steps taken in the method we developed to support our work [10]. Earlier in the process, our methodology was less structured and more ad hoc, creating challenges when non-analysts wanted to know what was going on or see the value of the work. Thus, the process was created and used to help translate and transfer knowledge, making explicit what was happening, why, and how others could use the results when done.

We show how we applied those steps for clustering players in *My Singing Monsters*. For this chapter, we will not describe the technical details but instead focus on the interpretations, written to be digestible by those without a technical background. A game analyst might include the contents of this chapter in a technical report or a presentation for other game analysts, game producers, or shareholders. For further technical details, including a study of two other, more complex models, please refer to [10].[1]

2 A Method for Game Analytics

As noted above, to support our journeys in game analytics, we created a five-step method for completing any analysis to assist game analysts with game analytics. At each step, it is essential to consider how to express what was completed since this work would typically be presented later to others. Furthermore, the audience may not have technical knowledge, so it is crucial to communicate in a way that is suitable for the audience.

In this chapter, we illustrate the application of this method with an analysis of the game *My Singing Monsters*. In [10], we perform a comparative analysis of four models for clustering players. Here, we will focus on the data translation aspects of the process and instead compare a base model to an augmented one. We present the five steps of the method in Fig. 1.

Firstly, it is crucial to understand the game. Game analysts often overlook this step, but it is a good idea for game analysts to play through the game, even partially,

---

[1] For a more detailed exploration of the data, the interested reader can find the source code for the analysis here: https://github.com/tanjo3/wae-clustering.

Fig. 1 A process flowchart showing the method we detail in [10]. Each of these five steps is an important component in any game analysis

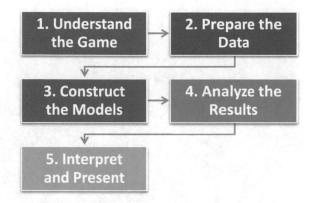

to understand the basic mechanics and begin to understand how players might feel while playing the game. This step might involve speaking with players, and one should include such interviews in the analysis.

Secondly, one needs to collect the data used in the analysis. Game studios often have pipelines for data collection, so this step involves carefully selecting data points from that data set. The first step is important here as it gives the context necessary to make good choices at this step. It is also important to consider biases in data selection here.

Thirdly, one needs to construct the models. In our work, this is the subject of the analysis where we are looking to compare different models' performances. However, for this chapter, we will quickly go over this section, focusing instead on the other parts of the analysis.

Fourthly, one applies the models constructed in step 3 to the data collected in step 2. Here, we want to look at the direct results from the analysis, such as the evaluation of the models used and identifying errors and anomalies. This is an important step as it justifies the interpretations made in the following step.

Finally, we want to provide interpretations of the results. This step is the most important as these will be the salient points in a report to interested parties. An excellent way to present things here is through visualizations from which the viewer can quickly glean information. The game analyst should accompany these with explanations written to be digestible by those without intimate knowledge of the techniques utilized in the previous four steps.

3 Understanding *My Singing Monsters*

My Singing Monsters is a multi-platform game available on PC/Steam, mobile, and PlayStation portable platforms. This chapter focuses on the game's mobile (iOS/Android) version, though they are functionally similar across platforms. Players can download *My Singing Monsters* for free and start playing. The game's basic

Fig. 2 Screenshot of My Singing Monsters, showing several monsters, structures, and decorations

premise is to breed singing monsters to populate the numerous in-game islands. The monsters also generate coins passively, which the player can collect to buy other monsters, decorations to decorate the island, or treats to level up the monsters. As the monsters level up, the rate at which they generate coins increases. Figure 2 shows a screenshot of Plant Island, along with various monsters and decorations.

To breed new monsters, the player must first select two monsters, and after a certain amount of real-world time, the monsters will produce an egg. The monster egg must then incubate for a certain amount of real-world time, after which it will hatch. The time these steps take, along with the chance the egg will be of a new monster type, changes depending on which two monsters are bred together. The player can speed up the breeding or incubation process using diamonds or watching advertisements in the game. The primary method of obtaining methods is by purchasing them with real-world currency.

There are, thus, multiple ways to play the game. Some example goals include breeding all varieties of monsters, generating coins as efficiently as possible, or making the islands look as beautiful as possible. Understanding how the players play the game and appreciating the diversity in how the game can be played is vital. For instance, if one way of playing the game leads to low player retention, it is worthwhile to investigate why that is the case and change the game to make that playstyle more enjoyable.

4 Preparing the Data

We now turn our attention to the data of *My Singing Monsters*. Data collection is an engineering effort on its own, and Big Blue Bubble, like many companies in the industry engaging in game analytics, has an entire division devoted to it. In general, the acquisition workflow depicted in Fig. 3 is followed, which matches general industry practice [12–14].

In the workflow shown in Fig. 3, data is ingested from game clients and servers using instrumentation embedded in their source code and brought into the analytics infrastructure through a web gateway service. From there, this data is streamed into storage, a combination of slower, high-capacity data stores for permanent data warehousing and high-efficiency databases for quick access to the subset of data needed for more immediate use. Data is then processed and analyzed to prepare it for presentation and dissemination within the organization. These last three workflow stages are executed continuously on a tiny portion of data to support near-real-time dashboards of critical measurements. At the same time, the bulk of the data is processed nightly for use the following day. This is how Big Blue Bubble's system functions today.

In the case of *My Singing Monsters*, following the game's terms of use, every player's action is recorded. Each player account has an associated identification number recorded along with each action to keep the data anonymized. Since we record every action, the amount of data generated daily is massive, which becomes a challenge when analyzing the data. A standard method is to aggregate the data. For example, we might gather all the events for a single day per player and use these as the input features for a data model. We might also combine various events into aggregate data points. Drachen et al. [5] took this approach and used six hand-crafted aggregate gameplay features to represent a player. They then used those features to identify clusters of player behaviors for the players of the game *Tomb Raider: Underworld*.

However, this method has some potential disadvantages. For one, selecting or creating good features requires time and domain expertise. Additionally, analysts may unintentionally impart human biases onto the data. For example, an analyst might exclude specific data points because they are considered unimportant. However, those data points may be helpful for certain analyses. Therefore, when choosing what data to work with, it is crucial to keep this in mind.

For our analysis in [10], we considered the features shown in Table 1 for each player. Each feature is an aggregate feature, and we choose to aggregate by day. Other analyses might aggregate by week, per player, or even look at individual events. We chose these features as it covers a broad range of player actions. We look

Fig. 3 The data acquisition workflow followed at Big Blue Bubble

Table 1 Daily features used
in player clustering

Player features	Gameplay features
Number of sessions	Minimum and maximum player level
Number of seconds played	Number of coins earned/spent
Number of ads watched	Number of diamonds earned/spent
Number of in-game purchases made	Number of treats earned/spent
Number of ad offers completed	Number of monsters bred/bought/sold

Note This table lists the features used by Tan and Katchabaw [10]
to represent players. Each of the features is a daily aggregate of
the related events. For instance, the number of coins spent refers
to the total number of coins the player spent that day

at each feature's value across the first 60 days since account creation. In summary,
we represent each player as a sequence of length 60 of 16-dimensional vectors.
Our analysis uses a random sample of approximately 100,000 *My Singing Monsters*
players.

5   Constructing the Models

Even if we want to use sequences to represent our players, traditional clustering
algorithms, such as those we will be using here, do not work with sequential inputs.
We use an *autoencoder* to solve this issue, a neural network model whose output
attempts to match its input. These models have a bottleneck, forcing them to "com-
press" the input. As such, we can use autoencoders to learn vector embeddings for
data. To have the autoencoder accept sequences for both input and output, we use
gated recurrent units (GRUs) [2], making our models recurrent. This terminology
comes from the neural network recursively applying its hidden units to each element
in the input sequence.

Our analysis in this chapter compares two different autoencoder models for player
clustering. The first is a basic recurrent autoencoder, the RAE, whose input and
output will be the sequential player representations we constructed. The second
model enhances the RAE by using the Wasserstein distance when evaluating the
similarity between the input and output. We call this called a recurrent Wasserstein
autoencoder or the WAE. This technique was first introduced by Tolstikhin et al.
[11]. Our analysis will evaluate whether the WAE clusters the players better than the
RAE.

5.1 Two Clustering Algorithms

There are multiple algorithms for clustering; our work in Tan and Katchabaw [10] used two different algorithms for our analysis. The main difference between the two is how the representative for each cluster is defined. Considering every single point in each cluster is unwieldy when looking at large data sets. Ideally, we would gain a general understanding of the cluster based on a single representative.

The first algorithm is the k-means algorithm, a traditional clustering technique that iteratively adjusts cluster membership until a stable state is found. The cluster representative is defined as the cluster's center of mass, referred to as the centroid. The centroid is the average of all values in that cluster.

The second algorithm used is archetypal analysis (AA) [3]. In contrast to k-means, the cluster representative is instead the most extreme member of the cluster. Effectively, the member with the most distinct values for features is chosen as the representative, referred to as the cluster archetype.

6 Analyzing the Results

This section discusses the results of applying the two models to the player data. Here, we want to determine the appropriate number of clusters to consider when looking at our players.

6.1 Clustering Metrics

One of the primary difficulties with clustering is how to evaluate it. Evaluation is straightforward when the correct cluster labels are known for the data set. However, since in our case, we do not know the players' cluster membership and the number of clusters in the data set, the metrics serve only as a heuristic. In our analysis [10], we evaluate the clustering using two well-known metrics: the Calinski-Harabasz index [1], where the higher the index value, the better, and the Davies-Bouldin index [4], where a lower index value is preferred.

6.2 k-Means Versus Archetypal Analysis

In Fig. 4, we plot the results of the clusterings. The left plot shows the clusterings evaluated using the Calinski-Harabasz index. The right plot shows the clusterings evaluated using the Davies-Bouldin index. For both plots, the x-axis corresponds to the number of clusters used when applying the clustering algorithm. We show

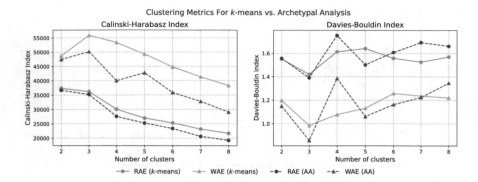

Fig. 4 Line graphs that plot the clustering metrics for *k*-means versus archetypal analysis while varying the number of clusters. We use solid lines for clusterings using *k*-means and dashed lines for archetypal analysis results. The left plot shows the Calinski-Harabasz index when varying the number of clusters formed, and a higher value is preferred. The right plot shows the Davies-Bouldin index for the clusterings, and a lower value is preferred. Figure adapted from [10]

the four combinations of autoencoder model and clustering algorithm in each plot. Results using *k*-means use a solid line, and results using archetypal analysis (AA) use a dashed line. We show the results for the RAE using circular markers and use triangular markers for the WAE. It is imperative to consider these visualization details when creating these plots, as the viewer should be quickly able to identify the information they need.

Recall that higher values for the Calinski-Harabasz index are preferred and that the opposite is true for the Davies-Bouldin index. Observing Fig. 4, we can see that the WAE outperforms the RAE in all cases. The results using the Calinski-Harabasz index are straightforward. The index values for the *k*-means clustering are always better than the corresponding values for AA. Additionally, the index value peaks at two clusters for the RAE and three for the WAE. For the Davies-Bouldin index, the best index value occurs when there are three clusters. Interestingly, the Davies-Bouldin index value at three clusters is better for AA than *k*-means. These results imply that using the WAE to cluster players into three clusters is the best option.

7   Interpreting the Analysis

In this section, we apply step 5 of our method and translate the clusterings into actionable interpretations. Doing so is a critical part of the process, where we can genuinely impart the value of our results to others. We do this using visualizations that let us quickly identify possible relationships within the data. First, we need to lower the dimensionality of the data to two (or three) dimensions. Then, we can color the data points representing players depending on which features we are trying to visualize. Finally, we can summarize the key points of each analysis. The section can constitute a report that might be presented internally within a game studio or even externally to shareholders or other interested non-technical parties. For other examples of visualizing and presenting data for game analytics, see El-Nasr et al. [7].

7.1 Visualizing High-Dimensional Data

To visualize high-dimensional data and the clusterings, we employ a dimension reduction technique in [10] called UMAP [9]. UMAP constructs a high-dimensional graph for the data and then creates a low-dimensional graph with a similar topology. We use UMAP over another well-known dimension reduction technique called t-SNE [15] since UMAP is faster and separates the clusters better. To display our visualization, we will use a popular plotting library for Python called *Matplotlib* [8].

7.2 Cluster Visualization

In Fig. 5, we use UMAP to plot each player in the data set, color-coded by cluster. We will consider three clusters. For consistency, we number the clusters in descending order of their sizes. We also choose the color for each cluster in the same way. We show the clusterings produced by k-means and mark their respective centroids with a star on the figure's left. We show the clusterings produced by archetypal analysis and mark their respective archetypes with a cross on the right.

We notice that the largest cluster typically contains strands of players. On the other hand, the smallest cluster is more compact. Since UMAP attempts to preserve the global structure of the data, we can interpret this as meaning that there is a more significant variance in behaviors in the larger clusters. Additionally, there is much overlap between the clusters between the two algorithms. For instance, the membership in the blue clusters is similar. In contrast, archetypal analysis puts slightly fewer players in the compact smallest cluster. We can observe this by seeing the smaller size of the red cluster for the WAE model when using AA. Overall, this might suggest that we can use both algorithms to produce similar clusterings but with different cluster representatives.

7.3 Day 120 LTVs

In Fig. 6, adapted from [10], we overlay each player's day-120 lifetime value (D120 LTV), which is the total revenue they have generated after playing for 120 days, on top of the UMAP visualization. Additionally, we log-transform the D120 LTV to

Player Clusterings

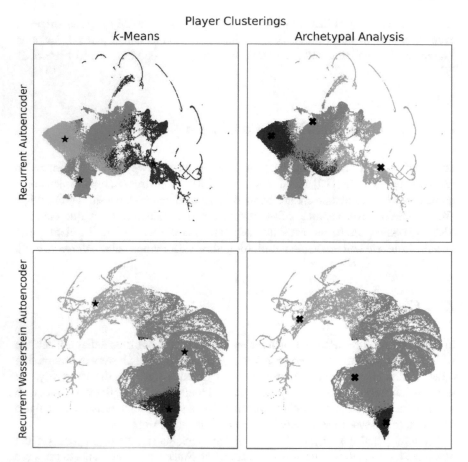

Fig. 5 Scatter plots showing the player clusterings. Each column represents a clustering algorithm (either k-means or archetypal analysis), and each row represents an autoencoder model. Each point represents a player, colored by their cluster membership. We color the clusters by size to keep colors consistent throughout all eight clusterings. For k-means, we mark the cluster centroid with a black star. For archetypal analysis, we mark the cluster archetype with a black cross. Figure adapted from [10]

deal with the wide range of values the D120 LTV takes. To account for zeros, we add one to each LTV before applying the transformation.

These plots show some correlation between D120 LTV and how the autoencoder models have clustered the players. Players in the denser areas of the clustering have higher LTVs, while players in the sparser areas tend to have lower LTVs. Now that we have evidence of this correlation, we can investigate these clusters to see what behaviors correlate with high or low LTV.

Day-120 LTVs

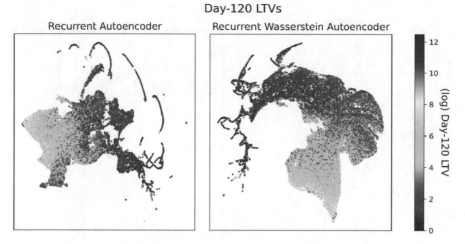

Fig. 6 Scatter plots showing the players' D120 LTVs. Each point is a player, as represented by both autoencoder models and colored by their log-transformed D120 LTVs. The color bar on the right of the figure shows the coloring, with blue representing a lower LTV and right representing a higher LTV. Comparing this figure with Fig. 5, we see that players with the highest LTVs tend to be in the smallest cluster. Figure adapted from [10]

7.4 Acquisition Source

Game studios will often run user acquisition (UA) campaigns with ad companies to bring people into the game. These campaigns are often expensive to run and maintain. We consider players that play the game via these campaigns to have been acquired non-organically. Figure 7, also adapted from [10], plots whether a player was acquired non-organically. In contrast to Fig. 6, we observe no correlation between the player's acquisition source and how the models have clustered them. This implies that the game features we used are not indicators of whether a player is organic. If there were correlations, we could investigate the UA campaigns' details and try to understand why they did or did not work. Overall, such an investigation could save the game company thousands of dollars.

7.5 Representative Analysis

Finally, we briefly examine some of the cluster representatives' daily features to compare and contrast k-means and archetypal analysis. We only look at six player features in this chapter and only consider the WAE model since it outperforms the RAE. In Fig. 8, we show daily features for the k-means clustering, and in Fig. 9, we show the same features for the archetypal analysis clustering. In both figures, we consider three clusters.

Acquisition Source

Recurrent Autoencoder   Recurrent Wasserstein Autoencoder

● non-organic   ● organic

Fig. 7 Scatter plots showing the players' acquisition sources. Each point is a player, and there is one plot for both autoencoder models. We see that most players are organic (not introduced to the game via a user acquisition (UA) campaign), so much so that it is hard to see non-organic players in the plots. Similarly, we see that the distribution of organic players is uniform, suggesting no correlation between the acquisition source and how the models clustered the players. Figure adapted from [10]

A noticeable feature is how much smoother the lines in Fig. 8 are compared to those in Fig. 9. This directly results from the fact that k-means is an average over all the players in a cluster. As a result, the features take on values that may be invalid. For instance, in Fig. 8, observe how the number of sessions for clusters 1 and 2 takes on a fractional value after ten days. We infer, however, that, on average, players in cluster 1 tend to be more active than in cluster 2. Since clusters are numbered in descending order of their sizes, this means that the largest and smallest clusters contain more active players than cluster 2. This is also reflected in the archetypal analysis clustering in Fig. 9. For this plot, we can infer another result in that players in cluster 1, while significantly less active than players in cluster 3, seem to log in for brief periods over the first 60 days. On the other hand, the cluster archetype for cluster 2 only logs in once.

Another thing we can observe in the plots in Fig. 8 is how cluster 2 has a good amount of activity at first but then seems to lose interest over time quickly. Figure 9 gives us more insight as we see the cluster archetype for cluster 1 logs in, plays, and watches ads more at this early stage than the archetype for cluster 3, which ends up more active at day 60. This might imply that they end up hitting a bottleneck which causes them to lose interest. A further investigation is required, but the wait times for breeding and incubation could be a factor. Thus, we might recommend that the game producer looks into ways to make other parts of the game more interesting. Such actionable suggestions are essential to convey to shareholders, and plain language

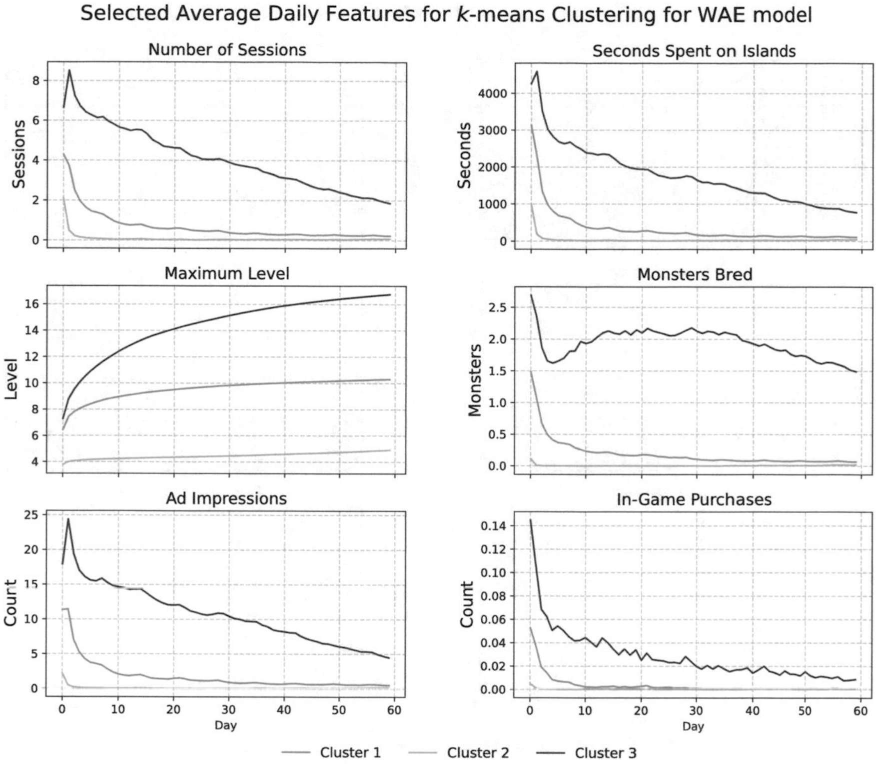

Selected Average Daily Features for k-means Clustering for WAE model

Fig. 8 Line plots showing a selection of average daily features for k-means clustering for the WAE model. Note that these values are averages over all players in the cluster

and interpretable graphs help all technical and non-technical parties better understand the players.

This analysis shows the value of k-means to get a big-picture view of a cluster. However, using both clustering methods to view players through multiple lenses is the ideal use case. Since the cluster representatives for archetypal analysis are actual players, the player's behavior can be more directly identified.

8 Concluding Remarks

This chapter presents a case study in game analytics using *My Singing Monsters*, taking us on a journey through our previous work in [10], providing additional insight and behind-the-scenes retrospectives to help others interested in this discipline. This analysis outlines a reusable methodology for game analytics that would-be game analysts could adapt and follow with a game of their choice. In particular, we leverage this methodology for player clustering and demonstrate how to describe the actions

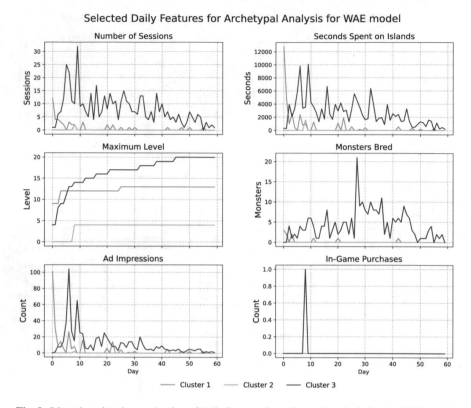

Fig. 9 Line plots showing a selection of daily features for archetypal analysis for the WAE model. Note that these are the features of actual players, restricting the values to legal values

taken at each step. We observe a correlation between the clustering and the players' D120 LTVs, but not their organic status. We also acknowledge how archetypal analysis might be a more interpretable clustering algorithm than k-means. However, using both algorithms in conjunction is the best to view player behavior from multiple angles.

However, this might raise concerns about the ethics of analyzing player data. Therefore, we keep the data anonymized to the best of our ability. Clustering players aims to identify what types of playstyles are present in the player base and to improve the overall quality of experience for all players.

In conclusion, we demonstrate in this chapter how to effectively communicate clusters of player behavior. Clustering is an important tool for game analysts, but one whose usefulness is often blunted by failing to follow a good process that culminates in effective translation and knowledge transfer. Visualizing the data makes it significantly easier for data translators to present an interpretation of the data to non-technical parties. Equipped with these interpretations, actionable suggestions, such as stopping a marketing or acquisition campaign or starting an in-game event to promote activity, can be made to game producers to benefit the game studio and its players. In the realm of game development, it is truly a win–win for everyone.

References

1. Calinski, T., Harabasz, J.: A dendrite method for cluster analysis. Commun. Stat. Theory Methods **3**(1), 1–27 (1974). https://doi.org/10.1080/03610927408827101
2. Cho, K., van Merrienboer, B., Gulcehre, C., Bahdanau, D., Bougares, F., Schwenk, H., Bengio, Y.: Learning phrase representations using RNN encoder–decoder for statistical machine translation. In: Proceedings of the 2014 Conference on Empirical Methods in Natural Language Processing (EMNLP), pp. 1724–1734 (2014). https://doi.org/10.3115/v1/D14-1179
3. Cutler, A., Breiman, L.: Archetypal analysis. Technometrics **36**(4), 338–347 (1994). https://doi.org/10.1080/00401706.1994.10485840
4. Davies, D.L., Bouldin, D.W.: A cluster separation measure. IEEE Trans. Pattern Anal. Mach. Intell. PAMI **1**(2), 224–227 (1979). https://doi.org/10.1109/TPAMI.1979.4766909
5. Drachen, A., Canossa, A., Yannakakis, G.N.: Player modeling using self-organization in Tomb Raider: underworld. In: 2009 IEEE symposium on computational intelligence and games, pp. 1–8 (2009). https://doi.org/10.1109/CIG.2009.5286500
6. Drachen, A.: DIREC TALKS: changing the Game: How Data Science has Transformed the Games Industry [Video]. DIREC Talk. https://direc.dk/direc-talks-game-data-science/ (2022)
7. El-Nasr, M.S., Drachen, A., Canossa, A. (eds.): Game Analytics: Maximizing the Value of Player Data. Springer (2013)
8. Hunter, J.D.: Matplotlib: a 2D graphics environment. Comput. Sci. Eng. **9**(3), 90–95 (2007). https://doi.org/10.1109/MCSE.2007.55
9. McInnes, L., Healy, J., Saul, N., Großberger, L.: UMAP: uniform manifold approximation and projection. J. Open Source Softw. **3**(29), 861 (2018). https://doi.org/10.21105/joss.00861
10. Tan, J., Katchabaw, M.: A reusable methodology for player clustering using Wasserstein autoencoders. In: Göbl, B., van der Spek, E., Baalsrud Hauge, J., McCall, R (eds.) Entertainment Computing—ICEC 2022, vol. 13477, pp. 296–308. Springer International Publishing (2022). https://doi.org/10.1007/978-3-031-20212-4_24
11. Tolstikhin, I., Bousquet, O., Gelly, S., Schoelkopf, B.: Wasserstein auto-encoders. http://arxiv.org/abs/1711.01558 (2019)
12. Weber, B.: A fully-managed game analytics pipeline. Game Developer. https://www.gamasutra.com/blogs/BenWeber/20180401/315962/A_FullyManaged_Game_Analytics_Pipeline.php (2018a)
13. Weber, B.: A history of game analytics platforms. Game Developer. https://www.gamasutra.com/blogs/BenWeber/20180409/316273/A_History_of_Game_Analytics_Platforms.php (2018b)
14. Wiger, N.: Connecting with Your Customers—Building Successful Mobile Games through the Power of AWS Analytics [Video]. GDC Vault. https://www.gdcvault.com/play/1021876/Connecting-with-Your-Customers-Buildin (2015)
15. van der Maaten, L., Hinton, G.: Visualizing data using t-SNE. J. Mach. Learn. Res. **9**(86), 2579–2605 (2008)

Language Corpora and Principal Components Analysis

Leslie Redmond, Denis Foucambert, and Lucie Libersan

Abstract The increase in use of statistical analyses to represent linguistic data constitutes an important turning point in the field of linguistics, allowing researchers to move beyond more traditional intuitive and observational paradigms. Language corpora are particularly difficult to analyze using inferential statistics and their analysis is much better handled through exploratory statistical analyses. This chapter will offer a brief overview of the use of statistics in the field of linguistics, focusing specifically on challenges associated with analyzing language corpora. We will then show how a Principal Components Analysis can be used to analyze a dataset of authentic texts of different genres written by post-secondary students in Quebec. This analysis was transformed into teaching modules designed to help post-secondary students improve their writing skills and to provide post-secondary instructors with an empirically-based framework to teach and evaluate genre-specific writing skills in French.

Keywords Language corpora · Exploratory analysis · Principal components analysis · Writing skills · French

L. Redmond (✉)
Memorial University of Newfoundland, St. John's, NL, Canada
e-mail: leslie.redmond@mun.ca

D. Foucambert
Université du Québec à Montréal, Montreal, QC, Canada
e-mail: foucambert.denis@uqam.ca

L. Libersan
CEGEP Ahuntsic, Montreal, Canada
e-mail: Lucie.Libersan@collegeahuntsic.qc.ca

© The Author(s), under exclusive license to Springer Nature Switzerland AG 2023
D. G. Woolford et al. (eds.), *Applied Data Science*, Studies in Big Data 125,
https://doi.org/10.1007/978-3-031-29937-7_9

1   Introduction

The increase in use of statistical analyses to represent linguistic data constitutes an important turning point in the field of linguistics, allowing researchers to move beyond more traditional intuitive and observational paradigms. The trend of increasingly robust and complex classical inferential statistical analyses opens opportunities to ameliorate and to reflect on how statistical analysis and data can continue to inform the field of linguistics (e.g., [16, 17]; Loewen and Gass [23]). The results of these more complex analyses can potentially better inform cognate areas and professional practice in areas such as language assessment, literacy, and language pedagogy.

This chapter will offer a brief overview of the use of statistics in the field of linguistics, focusing specifically on challenges associated with analyzing language corpora. We will then show how a Principal Components Analysis (PCA) can be used to analyze a dataset of authentic texts of different genres written by post-secondary students in Quebec. This analysis was transformed into teaching modules designed to help post-secondary students improve their writing skills and to provide post-secondary instructors with an empirically-based framework to teach and evaluate genre-specific writing skills in French.

2   Statistical Analyses in Linguistics

While there is a growing tendency to rely on more and more complex statistical analyses (e.g., multilevel models) in the field of linguistics, researchers note that there tends to be relatively few statistical procedures used in linguistics research and that the chosen analyses are not always adequate for modelling complex relationships [27, pp. 3–4]. Training in statistics for graduate students and researchers in the humanities and social sciences tends to be primarily based on the discovery and mastery of descriptive and inferential methods. While these analyses provide an adequate framework to capture statistical relationships in experimental studies, these types of analyses often fail to capture the nuance and depth of data in linguistic corpora.

There is, however, another major trend in statistics known as the *French School of Data Analysis* (e.g., [1]). The principles of this type of statistics are based on the work popularized by Benzecri [3]: statistics without probability or an a priori model. In this case, "the model must follow the data, not the other way around" (our translation of Benzecri et al. [4, p. 6]). Benzecri opposes the idealist approach of Noam Chomsky who, in the 1960s, defended the necessity of a priori modelling that would provide a framework for understanding linguistic structures. On the contrary, Benzecri [2] proposes an inductive method of analyzing linguistic data in line with the goals of the distributionalists [5, 10] which were to construct the rules of grammar from actual corpora [1].

2.1 Exploratory Statistical Analyses in Linguistic Research

The use of exploratory statistical modelling through different types of factor analysis to analyze corpus data is growing (e.g., [6, 18, 24, 27]). Generally, factor analyses are used to reveal the underlying structure of data and are particularly applicable to large data sets where measured variables can be reduced to latent factors. The type of factor analysis chosen depends on the type of initial variables collected by the researcher. PCA describes a set of numerical variables, simultaneously accounting for variance both within and across variables [12, 25]. This type of analysis will be the focus of this chapter.

Linguistic research, and in particular corpus-based research, often faces the difficulty of having to deal with large numbers of variables simultaneously. Moreover, these variables are often correlated. For example, sentence length is necessarily related to the number of elements it contains and that the more adjectives there are, the more nouns there are. To avoid this problem of multicollinearity or to reduce the number of dimensions that the researcher must analyze, exploratory statistical analyses are particularly useful.

2.2 Principal Components Analysis

One of the goals of a PCA is to reduce the dimensionality of a dataset, while preserving as much 'variability' (i.e., statistical information) as possible (e.g., []). In other words, the objective is to reduce the data by taking the measured or observed variables and turning them into latent factors or components while still being able to account for the variance within and across variables. The researcher starts with a table containing a large number of variables and observations. It is impossible for the naked eye to spot regular elements in this magma of data. The objective of PCA is to replace this initial table, which is difficult to read and above all impossible to represent graphically, with tables that are easier to read and with understandable graphical representations, while losing as little information as possible.

The new tables consist of new variables that are constructed from linear combinations of the initial variables. These new variables are the principal components of PCA and are referred to as factors, axes, or dimensions depending on the discipline. The analysis creates as many combinations as there are variables, and within a classical conception of PCA, these principal components are orthogonal. This means that they are uncorrelated and therefore represent information which is strictly different from all the other components [12, pp. 420, 425].

Once these components have been created by the PCA, the researcher's job is to reflect on the meaning of each of these new variables, the components. After deciding on the number of components to observe (methods for doing this will be discussed further in this chapter), they must interpret the variables that contribute to each of the selected components. To do this, the researcher has several interpretation aids at their disposal, as we shall see later in this chapter.

2.3   Current Study

The data presented in this chapter are part of a larger study which looked at context-dependent writing strategies across disciplines in CÉGEPS in Québec[1] [19, 20]. The original study and analysis [19, 20] analyzed the linguistic and textual features of 14 different types of texts using 72 variables. The overarching goal of this work was to provide specific guidance to college professors around the textual characteristics of different genres to provide them with specific tools to teach and assess genre-specific writing skills. These guidelines (see [19]) as well as a corresponding research report (see [21]) were published by the *Centre collegial de développement de matériel didactique (CCDMD)*. Each textual genre has a corresponding module which is broken down into evidence-based pedagogical material focusing on the general nature of the genre, writing challenges and strategies, revision strategies and corresponding pedagogical activities. There is also a teacher's guide which focuses more specifically on language structures which need to be mastered for each of the textual genres.

For the purposes of this chapter, we will demonstrate the data analysis procedure that led to this innovative and publicly accessible pedagogical material on three genres of texts (business letters, CVs and argumentative essays) and 29 linguistic variables. These variables were measured using the language correction software Antidote 9 [7] and then systematically verified by the researchers. The number of occurrences of each of the variables was converted into a percentage using the number of words as the denominator.

The choice to demonstrate the analysis on a subset of our data is pedagogical in nature as the results, certainly with regard to data visualization, are easier to interpret and explain.

The retained linguistic variables are presented in Table 1.

3   Analysis and Interpretation

Using the data set described above (29 variables which are represented by the percentage of occurrences), we will take readers through the steps of a PCA on the subset of our data, discussing implications for interpreting and translating the data throughout. This analysis will be demonstrated by using *jamovi* [29], a free and open statistical platform which is continuously updated by a community of users.

---

[1] CEGEPs are publicly funded colleges which students in Quebec, Canada attend after high school. They offer a variety of pre-university, vocational, and technical programs.

Table 1 Variables analyzed (adapted from [20])

Category	Variable
Indicators of presence or absence of interlocutors	1st person pronouns (singular and plural)
	2nd person pronouns (singular and plural)
	"On" pronoun
Indicators of time and place	Number of verbs in the infinitive
	Indicative: present
	Indicative: imparfait
	Indicative: future
	Indicative: passé composé
	Indicators of time (adverbs, etc.)
	Indicators of place (adverbs, etc.)
Indicators of modality	Auxiliary "pouvoir"
	Modal auxiliary
	Passive
	Negation
	Impersonal
	Subjunctive
	Conditional
	Imperative
	Enunciation modalities
Indicators of syntactic complexity	Number of conjugated verbs
	Number of words
	Number of sentences
	Number of long sentences
	Relative pronouns
	Prepositions
	Coordinating conjunctions
	Subordinating conjunction
	Anaphora
	Logical connectors

3.1 Verify Assumptions

The first step in conducting PCA is to verify the underlying assumptions of PCA. These assumptions include the nature of the variables (continuous) and relationships between them (linear), sampling adequacy and the nature of the data (suitable for reduction) (e.g., Field [8]; [22]). Students and researchers should be able to justify their choice of analysis based on the nature of the data they are analyzing and the assumptions of the chosen test, in this case, PCA.

3.1.1   Variables

PCA requires variables that are continuous and that have linear relationships between them. The 29 selected variables are continuous variables (% of occurrences). There are multiple ways to verify the assumption of linear relationships: correlation matrices, or visual inspection of data using scatterplots, for example. Students generally have a good grasp on the nature of different types of variables, selecting variables, and the concept of linear relationships between continuous variables.

3.1.2   Sampling Adequacy and Suitability of Data for Reduction

There are two main statistical measures used to verify sampling adequacy and the suitability of data for reduction: The Kaiser–Meyer–Olkin (KMO) test of sampling adequacy and Bartlett's test of sphericity.

KMO Test of Sampling Adequacy

The Kaiser–Meyer–Olkin (KMO) test of sampling adequacy provides an indication of how appropriate a sample is for a specific analysis. The value, which is between 0 and 1, "[…] represents the ratio of the squared correlation between variables to the squared partial correlation between variables" [8, p. 647]. *Jamovi* calculates this statistic for both individual and multiple variables. According to Kaiser [14], 0.50 is the lowest acceptable measure and scores above 0.7 are considered good. The higher the score, the more adequate the sampling. The overall KMO statistic for this sample is 0.725, falling between "middling" and "meritorious" according to Kaiser ([14], p. 35) and thus acceptable for PCA.

Bartlett's Test of Sphericity

PCA is carried out on continuous variables with linear relationships, and thus, a correlation matrix underlies the PCA analysis. Broadly, Bartlett's test of sphericity tests the null hypothesis that the underlying correlation matrix is not significantly different than an identity matrix (a matrix of 1s on the diagonals and 0s elsewhere) [8, p. 589]. When Bartlett's test is significant, it indicates that data is significantly different than an identity matrix and thus suitable for data reduction. Tests for sphericity are common in many inferential statistical analyses which require the verification of the assumption of equal variances.

3.2 Conduct Analysis

Given that all assumptions have been met: our variables are continuous and have a linear relationship, our sampling is adequate, and the data is suitable for reduction, we can now proceed with the PCA. First, we will discuss criteria for factor retention, followed by factor loadings and rotation. Then, we will discuss options for data visualization. Throughout, we will pay particular attention to areas of difficulties for students especially with regard to the interpretation of the analysis and how this informs data translation.

3.2.1 Criteria for Retention of Factors (Components)

There are multiple criteria that can be used to decide how many factors should be retained in the final analysis. PCA requires that the researcher makes informed decisions based on the research surrounding factor retention. We will discuss different techniques used for factor retention: eigenvalues, scree plots, and parallel analysis.

Eigenvalues are scores which represent the amount of variance in each of the extracted components or factors (e.g., [22]). The higher the eigenvalue, the more variance it accounts for in the components. Eigenvalues can be a useful tool to identify which factors to retain. A common rule of thumb for PCA is Kaiser's [13] criterion which states that components for which the Eigenvalue is greater than one should be retained (e.g., [8, 22]).

Scree plots can also be used as an indicator of how many factors to retain. Scree plots graph eigenvalues in descending order, providing a visual representation of the relative explanatory power of each of the extracted components. The visual examination of a scree plot is somewhat subjective and different researchers may come to different conclusions. Figure 1 shows the scree plot associated with our data.

Visual examination of the scree plot would lead us to retain either the first three or five components, depending on where the point of inflection judged to be.

Finally, a less utilized, but more reliable [8] option for factor extraction and retention is *parallel analysis*. Parallel analysis is a technique whereby the eigenvalues generated from the measured variables are compared to that of a corresponding eigenvalue from a randomly generated dataset, comparable to the dataset being used (e.g., Horn [9,11]). Components are retained when their eigenvalues are bigger than their analogs from the randomly generated data set. Table 2 shows the Eigenvalues (SS loadings) and the percentage of variance explained by the three components retained by parallel analysis.

The scree plot in Fig. 2 compares the actual data with the simulations, indicating the point at which they can no longer be disentangled from each other.

The choice of how many factors to retain is one of the most important decisions a researcher will make when using PCA as these are the components that will need to be visualized, interpreted, and ultimately translated. It is imperative that students are explicitly taught how to make these essential decisions, which do not always

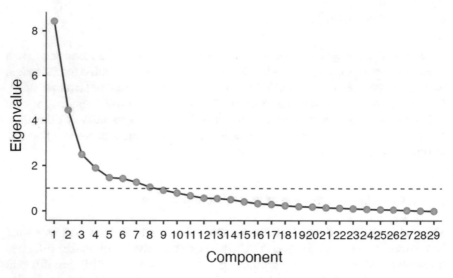

Fig. 1 Scree plot

Table 2 Components
retained through parallel
analysis

Summary			
Component	SS loadings	% of variance	Cumulative %
1	8.33	28.74	28.7
2	4.55	15.69	44.4
3	2.50	8.64	53.1

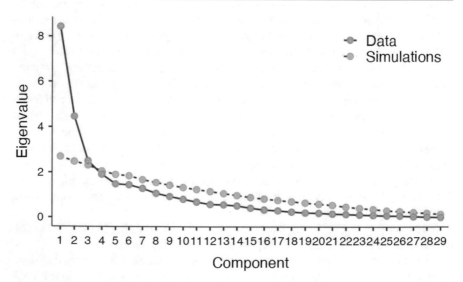

Fig. 2 Scree plot comparing data with simulations

have rigorous solutions. When teaching students how to interpret scree plots, and more broadly to choose how many components to retain, it is important to discuss the subjective nature of this choice.

In practice, the applicable criteria are empirical, such as the ones discussed above. However, other factors also drive this choice. These include a desire for model parsimony, the examination of candidate principal components for further analyses to see if they have any natural interpretation, and model selection by comparing models with differing numbers of components either through goodness of fit or through predictive accuracy on independent testing data.

Retaining components corresponding to eigenvalues greater than 1 (or by using scree plots, or parallel analysis) means that we are only interested in the components that contribute more explanatory power than one of the initial variables. In so doing, we lose a significant part of the total variance and therefore of the explanatory quality of the overall phenomenon that we are trying to explain. In our example, we would have had to consider and explain 8 components (following the eigenvalue criteria of greater than 1), which remains a difficult exercise since, by definition, we must respect the independence of the factors in the interpretation. An understanding of the different types of tools available to guide in the choice of the numbers of factors to retain is imperative for students. To this end, following Field [8], we retained a three-factor solution as determined by parallel analysis which provides both a rigorous and empirically valid solution.

3.2.2 Rotation

In factor analysis, the initial, unrotated solution is difficult to interpret as discriminating between factors is challenging and thus rarely reported on. "...[F]actor rotation effectively rotates the factor axes such that variables are loaded maximally to only one factor" [8, p. 642].[2] We have chosen to use orthogonal rotation with the *Varimax* method given that one of the underlying statistical assumptions of PCA is an orthogonal relationship between the created components the most common method of orthogonal rotation being *varimax*.

3.2.3 Factor Loadings

In a PCA, each measured variable can load onto one or more of the extracted factors. The researcher then needs to set a minimum threshold for considering the contributions of the observed variables to the extracted components. There is however debate in the literature surrounding criteria for factor loadings, with some authors suggesting that the cut-off changes depending on sample size (e.g., [8, 22]). We set our factor loading cut-off criteria at 0.4 following Pett et al. [26]. In our case,

---

[2] See Loewen and Gonulal [22] and Field [8] for a discussion of orthogonal versus oblique rotation. Loewen and Gonulal [22] report specifically on language data.

this allows to account for the nature of linguistic variables which will invariably contribute to multiple textual genres, but the importance of their contribution may differ.

In sum, PCA requires a certain number of decisions be taken by the researchers at each juncture of analysis (e.g., variable selection, factor retention, rotation, factor loading, etc.). These decisions are both theoretically (driven by considerations relative to the field of study) and statistically motivated. Once these decisions have been made, the data can be visualized, interpreted and conclusions can be drawn.

3.2.4   Data Visualization and Interpretation

To gain access to options for data visualization in *jamovi*, it is necessary to download the *snowCluster* module [28] from the *jamovi* library. We will explore two options for data visualization of a PCA analysis: PCA plot and Group plot.[3] We will also discuss the implications of these visualizations on the information that students and researchers can use for data translation.

PCA Plot

Initially, using the PCA plot function of the *snowCluster* module [28], we can project the variables which load onto the retained components and their relative contributions to the axes. This is shown in Fig. 3.

This plot shows the contributions of each of the measured variables to the components. At this stage, the figure is difficult to interpret, and it does not readily allow the information to be clearly understood. However, we can see that certain linguistic features are clustering together around the different axes. In the next step, in Fig. 4, we add on the individuals (the texts), allowing the visualization of both the measured variables and the texts from which they were measured.

The first component, on the horizontal axis, captures 28.7% of the variance. On the righthand side (positive contributions—Figs. 3 and 4), the variables capture long (number of words and number of sentences) texts which are also complex as they contain a large number of subordinating conjunctions and relative pronouns. There are also many negative propositions and anaphora (pronouns, etc., that refer to ideas introduced previously). The texts situated on the positive side of this first factor are thus relatively complex and use several linguistic means to achieve complexity and sophistication. The opposing negative side of the component is comprised of a significant number of indicators of time and place as well as verbs which are in the infinitive. The further a text is positioned on the negative side of the first factor, the less complex it is. These texts are brief and factual and do not exploit complex syntactic structures.

---

[3] Visualization of other dimensions (beyond the first and second) can be carried out through the module MEDA available in the *jamovi library.*

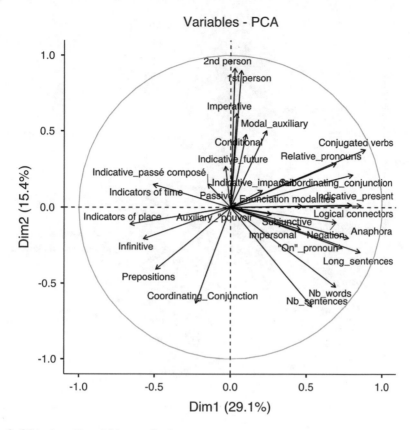

Fig. 3 PCA plot with variable contributions

Variables which load positively on the second component (represented on the vertical axis), indicate the presence of different interlocutors (first and second person pronouns) and the presence of different modes: orders (indicated by verbs in the imperative and circumspection (indicated by the conditional and modal auxiliaries). Texts which score highly on the second component demonstrate a relationship between interlocutors and with interaction between them that oscillates between injunction and negotiation. Conversely, texts with negative scores on this factor have simple syntactic constructions using mainly prepositions or coordinating conjunctions.

The third component is comprised of texts which are characterized by the use of the past tense, the passive voice and indicators of time. These texts are also characterized by the absence of the future tense, infinitives, and the auxiliary verb "pouvoir" (*can*).

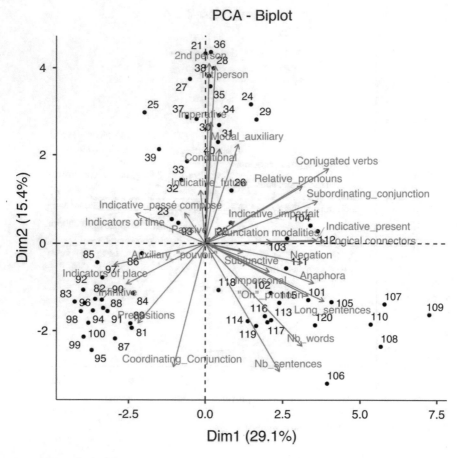

Fig. 4 PCA bi-plot with variable contributions and individual texts

Group Plot

In this final step of data visualization, we can now project a variable that we have not yet included in our analysis (the genre of the text) onto the PCA plot. This is also possible using the snowCluster module. Figure 5 shows the resulting image.

This graph now allows us to visualize the data in relation to the textual genres studied. The argumentative essay is located on the positive side (the blue square) of the horizontal axis and characterized by a certain level of complexity and sophistication in the writing. This blue square, representing the center of gravity of the argumentative essays, is also positioned on the negative side of factor 2 (vertical axis), showing that these texts do not present marks of the interlocutors engaged in a dialogue. The business letter texts are positioned on the positive side of factor 2, and the thicker red dot representing the center of gravity of this group is of zero value on factor 1. This means that that these two types of texts do not share any common features in terms of

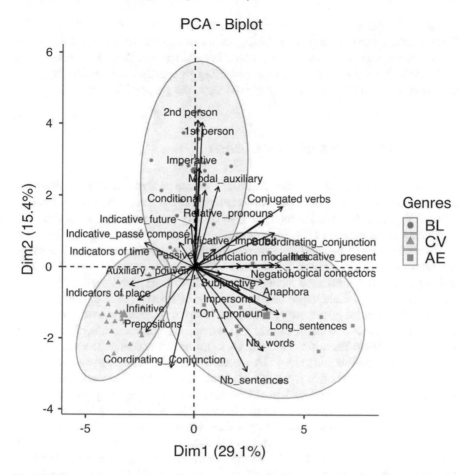

Fig. 5 PCA plot with variable contributions and genres of texts. *Note* CV = Curriculum vitae; BL = Business letter; AE = Argumentative essay

writing sophistication and complexity. Observing the elongated red ellipsis corroborates this interpretation. These texts are characterized by the manifestation of a significant dialogue—with its different modalities—between interlocutors explicitly present in the text. The same sort of reasoning could be applied to the final category, CV, characterized by the use of verbs in the infinitive, prepositions, coordinating conjunctions, and indicators of place.

Within each category, there is some variation in the position of the texts on the factorial axes as shown in Fig. 4. For example, one text (118) is very close to the vertical axis (factor 2). This text is relatively far from the center of gravity of its group and does not seem to have the same characteristics as the others. On the other hand, texts 107, 108 and 109 are positioned at the end of the horizontal axis (factor 1) indicating a particularly sophisticated and complex writing style, characteristic of argumentative essays.

At this stage, PCA allows the researcher to return to the characteristics of the individuals (the text, in this case) allowing for a more qualitative and precise analysis of a particular text, relative to the characteristics of the group. This allows researchers to characterize the textual categories according to specific, measurable linguistic characteristics which cluster around components, the genre of the text.

4 Data Translation and Conclusion

The data analyzed here was translated into data-driven pedagogical material designed to support students and college professors in the teaching and assessment of different genres of written texts used in many post-secondary programs. Further applications pertaining to the creation of assessment rubrics can also be imagined.

This pedagogical material for students and college professors took the form of modules which used this data to inform instruction and assessment of context-specific writing skills. If we take the example of business letters (https://www.ccdmd.qc. ca/media/Genres_05Lalettreprofessionnelle.pdf), in the associated module, specific guidance is provided on the use of indicators of time, personal pronouns, as well as verb tenses and modes which lead to tactful and polite communication, characteristic of business letters. Then if we look at CVs (https://www.ccdmd.qc.ca/media/Gen res_03Lecurriculumvit.pdf), the PCA revealed that CVs tend to use indicators of time far more than the other types of texts analyzed. Further, they are characterized by short, simple sentences which use verbs in the infinitive (not conjugated) that are linked by coordinating conjunctions (and, or etc.). The resulting guidance for students shows how it is important to make sure that sentences follow a similar structure by consistently using either noun groups to describe experience (e.g., experienced cashier; devoted animal-shelter volunteer) or verb groups (e.g., worked as a cashier for 5 years, volunteered at an animal shelter). Finally, a check list for students is also provided.

Students and researchers in linguistics, and more broadly in the social sciences and humanities, stand to benefit from learning about the philosophy and methodology of this type of inductive data analysis through its inclusion in statistical training at the post-secondary level in addition to the classical methods of probabilistic inferential statistics. More generally, many undergraduate and graduate students in linguistics pursue careers in cognate fields such as education, speech language pathology and artificial intelligence. A solid understanding of factor analysis methods could lead to a more comprehensive understanding and use of linguistic data which informs these areas of practice.

References

1. Beaudouin, V.: Retour aux origines de la statistique textuelle: Benzécri et l'école française d'analyse des données. In: Mayaffre, D., Poudat, C., Vanni, L., Magri, V., Follette, P., Daire, C., Couessurel, F. (eds.) JADT 2016, pp. 17–27 (2016)
2. Benzecri, J.-P.: Pratique de l'analyse des données. Linguistique et lexicologie: Vol. Tome 3. Dunod (1981)
3. Benzecri, J.-P.: Histoire et Préhistoire de l'Analyse des Données. Bordas (1982)
4. Benzecri, J.-P., et al.: L'analyse des correspondances. Dunod (1973)
5. Bloomfield, L.: Language. The University of Chicago Press (1983)
6. Desagulier, G.: Corpus Linguistics and Statistics with R: Introduction to Quantitative Methods in Linguistics. Springer International Publishing AG, Cham (2017)
7. Druide informatique: Antidote 9 [Computer Software] (2016). https://www.antidote.info/en/
8. Field, A.: Discovering Statistics Using SPSS, 3rd edn. Sage Publications Inc. (2009)
9. Franklin, S.B., Gibson, D.J., Robertson, P.A., Pohlmann, J.T., Fralish, J.S.: Parallel analysis: a method for determining significant principal components. J. Veg. Sci. **6**(1), 99–106 (1995). https://doi.org/10.2307/3236261
10. Harris, Z.S.: Distributional structure. WORD **10**(2–3), 146–162 (1954). https://doi.org/10.1080/00437956.1954.11659520
11. Horn, J.L.: A rationale and test for the number of factors in factor analysis. Psychometrika 30(2), 179–185 (1965). https://doi.org/10.1007/BF02289447
12. Hotelling, H.: Analysis of a complex of statistical variables into principal components. J. Educ. Psychol. **24**(6), 417–441 (1933). https://doi.org/10.1037/h0071325
13. Kaiser, H.F.: The Application of Electronic Computers to Factor Analysis. Educational and Psychological Measurement 20(1), 141–151 (1960). https://doi.org/10.1177/001316446002000116
14. Kaiser, H.F.: An index of factorial simplicity. Psychometrika 39(1), 31–36 (1974). https://doi.org/10.1007/BF02291575
15. Larson-Hall, J.: A Guide to Doing Statistics in Second Language Research Using SPSS and R. Routledge, New York (2015)
16. Larson-Hall, J., Herrington, R.: Improving data analysis in second language acquisition by utilizing modern developments in applied statistics. Appl. Linguis. **31**, 368–390 (2009)
17. Larson-Hall, J., Plonsky, L.: Reporting and interpreting quantitative research findings: what gets reported and recommendations for the field. Lang. Learn. **65**, 127–159 (2015)
18. Larsson, T., Plonsky, L., Hancock, G.R.: On the benefits of structural equation modeling for corpus linguists. Corpus Linguist. Linguist. Theory (2020). https://doi.org/10.1515/cllt-2020-0051
19. Libersan, L.: Stratégies d'écriture dans la formation spécifique (2012). En ligne: www.ccdmd.qc.ca/fr/strategies_ecriture
20. Libersan, L., Foucambert, D.: Un modèle exploratoire d'analyse de données textuelles au service de la didactique de l'écrit dans les collèges québécois. In: Neveu, F., cMuni Toke, V., Blumenthal, P., Klingler, T., Ligas, P., Prévost, S., Teston-Bonnard, S. (eds.) Congrès Mondial de Linguistique Française—CMLF'12, pp. 307–323. Institut de Linguistique Française (2012)
21. Libersan, L., Claing, R., Foucambert, D.: Stratégies d'écriture dans la formation spécifique. Rapport 2009–2010. CCDMD/Collège Ahuntsic, Montréal (2010). En ligne: www.ccdmd.qc.ca/media/doc_theo_div_Rapport_Formation_specifique.pdf
22. Loewen, S., Gonulal, T.: Exploratory factor analysis and principal components analysis. In: Plonsky, L. (ed.) Advancing Quantitative Methods in Second Language Research, pp. 182–212. Routledge (2015)

23. Loewen, S., Gass, S.: The use of statistics in L2 acquisition research. Language Teaching 42(2), 181–196 (2009). https://doi.org/10.1017/S0261444808005624
24. Paquot, M., Plonsky, L.: Quantitative research methods and study quality in learner corpus research. Int. J. Learn. Corpus Res. **3**, 61–94 (2017). https://doi.org/10.1075/ijlcr.3.1.03paq
25. Pearson, K.: LIII. On lines and planes of closest fit to systems of points in space. Lond. Edinb. Dublin Philos. Mag. J. Sci. **2**(11), 559–572 (1901). https://doi.org/10.1080/14786440109462720
26. Pett, M.A., Lackey, N.R., Sullivan, J.J.: Making Sense of Factor Analysis: The Use of Factor Analysis for Instrument Development in Health Care Research. Sage (2003)
27. Plonsky, L. (ed.): Advancing Quantitative Methods in Second Language Research. Routledge, New York (2015)
28. Seol, H.: snowCluster: Cluster Analysis [jamovi module] (2020). https://github.com/hyunsooseol/snowCluster
29. The jamovi project: jamovi (Version 1.6) [Computer Software] (2021). https://www.jamovi.org

A Tutorial of Analyzing Accuracy in Conceptual Change

Lin Li

Abstract Accuracy data have been collected in conceptual change studies to show students' understanding of a scientific concept, such as the time of swing of a pendulum. In addition to reporting accuracy or error rates with descriptive statistical terms, they are also analyzed inferentially with the experimenter's off-shelf statistical procedure, the Analysis of Variance (ANOVA). However, the inherent dichotomy of the correct/incorrect binary responses renders such a practice inappropriate. By contrasting the results analyzed with the repeated-measures ANOVA and regression-based modelling, the advantages of using the latter framework for data with a binary response are illustrated, especially in studying the near-ceiling performance. The comparison is done using a simulated dataset of negative priming in conceptual change research and contrasting these two statistical modelling methods side-by-side. The logistic regression approach to accurate data is recommended given its importance as a widely used statistical modelling method in fields such as healthcare analysis, credit rating, social statistics and econometrics. Moreover, it is widely available as a component of most general-purpose commercial statistical packages.

Keywords Accuracy · Analysis of variance · Generalized linear mixed model · Logistic regression · Near-ceiling performance · Repeated measures

1 Introduction

In education research, quantitative data can take many forms, such as time or accuracy. In contrast to measuring learners' performance in milliseconds, it is common for education researchers to collect students' responses and mark them as right or wrong given a pre-defined theoretical position. Both teachers and researchers in education tend to draw definitive conclusions from analyzing such data, particularly

Supplementary Information The online version contains supplementary material available at [https://10.1007/978-3-031-29937-7_10].

L. Li (✉)
University of Windsor, Windsor, ON, Canada
e-mail: li81@uwindsor.ca

after seeing a statistically significant result, yet reporting significance or not based on p-values thresholds of 0.05 or 0.01 has been contested as an acceptable good practice [9, 14]. The unsuitability is of great concern when the data modelling method may not be appropriate.

In this chapter, I first introduce the quantitative conceptual change studies in science education. Next, a simulated dataset is described, followed by a comparison of two types of statistical data analyses: the Analysis of Variance (ANOVA) versus the Logit Modelling (namely, a logistic regression approach). For them, a repeated measures data structure is highlighted due to the experimental design and the nature of the data. Although researchers in education and psychology may be more familiar with the former, its underlying assumptions aren't satisfied given the underlying nature of the response data. Consequently, the latter (logistic regression) is more appropriate. The ANOVA-to-regression transition is illustrated through an analysis of the simulated data.

2   Quantitative Conceptual Change in Science Education

The phrase conceptual change was coined in the 1970s as an umbrella term for characterizing science teaching and learning. In scientific knowledge acquisition, conceptual change refers to a knowledge restructuring process that characterizes a learner's active information processing [11]. In the literature, the knowledge learning process has also been named as knowledge "tree switching" [12], reintegrating "Knowledge in Pieces" [3], re-assigning ontological categories [2], restructuring a theoretical framework [13], to name a few.

The most common research method in the field is using a survey or interview, which generates proposition-related data subject to further re-interpretation. In contrast to such a verbal description, researchers have also documented quantitative datasets; for example, Canadian researchers introduced the Negative Priming paradigm into conceptual change studies [10]. In this paradigm, they focused on non-verbal measures, such as human reaction times and the proportion of the participants' correct responses over a set of item-based trials. The experimental procedure was designed to activate the students' pre-instructional ideas with some visual patterns in a priming stage first, with its aftereffects on subsequent learning measured afterward. Due to the incompatibility between the students' alternative conceptions and the physicists' scientific ideas, some error-prone and slower responses were expected in the experimental condition relative to the control. The critical evidence to support the claim of a robust conceptual change effect was a higher error rate observed on one condition relative to its control.

In 2018, Vosniadou and her collaborators also analyzed their accuracy data to support a similar conclusion. Commonly, the accuracy data were submitted to the ANOVA procedure for showing strong evidence of (i.e., "significant") group mean-level differences. In doing so, the researchers have treated their accuracy data (a true score plus a random error component) as a continuous dependent variable being sampled from a normal distribution because the ANOVA is based on such an assumption.

However, the sources of variability may come from at least two types of sources in the negative priming paradigm: within-individual variability caused by task items and interindividual variability of correct responses over time. Mapping binary correct or incorrect responses onto a continuous scale is not straightforward. According to Golay et al. [4], accuracy is often defined between a maximum and a minimum boundary (100–0% correct). Consequently, the assumption of a continuous dependent variable that follows a normal distribution would not hold. Similarly, Jaeger [6] called for dropping ANOVA-based techniques for analyzing categorical data transformed or not. Although the inappropriateness concern is well known, education researchers still widely use it in conceptual change studies.

In this chapter, we contrast two approaches to analyzing accuracy data: repeated-measures ANOVAs (linear regression) and generalized linear mixed-effects modelling. A mixed effects model is chosen because of the unique feature of the within-subjects design—each participant can contribute multiple data points, and these points may be correlated with one another because they are generated by the same participant. Such design-induced correlation violates the independence assumption of ANOVA. One way to address such a concern is to choose a mixed effects model, allowing a researcher to systematically separate the item-level variability (within subjects) from the participant-level variability (within groups).

3 A Binomial Distribution of Binary Responses

The concerns regarding analyzing accuracy-based data can be addressed by considering the fact that the data are generated from binary responses. At the fundamental level, a binomial sampling distribution would theoretically match the nature of the data-generating mechanism. Specifically, a binomial distribution can be used as a model for the number of correct responses in a fixed number of independent trials where p represents the probability that a response may be correct on each experimental trial. In conceptual change studies, this implies that a learnt or changed conceptual structure is more likely to lead to more correct observable responses, whereas an unchanged one keeps such a possibility at the control level. Thus, the probability p of h correct responses in n independent experimental trials could be assumed to be given by the following binomial probability mass function:

$$f(h, n, p) = \binom{n}{h} p^h (1 - p)^{n-h}, h = 0, 1, 2, ..., n \tag{1}$$

In the above formulation, the probability of success is assumed fixed. However, it is often desirable to model this as a function of other predictors. When learning a scientific concept, a student may rely on using a pre-instructional but incorrect intuition or switching to a correct one. Such a change in the conceptual structure determines how likely the student would respond in a series of trials. When the correct responses as the dependent variable are considered for (dis) confirming an underlying conceptual change status, the responses can be characterized either as a varying continuous accuracy rate or a binary one with two discrete values. For the former, the ANOVA methods are often used to tell apart the conceptual item-level and the participant-level variances for a group means comparison. For the latter, another type of statistical analysis is preferred, such as logistic regression, which can estimate the probability of a student choosing a correct answer as a function of the conceptual change status.

There are two more essential characteristics of the logistic regression approach. First, it is built upon the notion of odds, defined as the ratio of the probability of an event occurring relative to its probability of non-occurrence. Second, as its name suggests, a logistic regression equation is expressed as the logarithm or natural logarithm of odds: ln(odds). In a general form, it is written as:

$$\ln(odds) = \ln\left(\frac{p(x)}{1 - p(x)}\right) = \alpha + \beta_1 x_1 + \beta_2 x_2 + \cdots + \beta_q x_q \tag{2}$$

where the parameter α represents the logarithm of the odds of the baseline condition and each coefficient parameter β_k represents the natural logarithm of the odds ratio for the kth predictor variable, x_k $(k = 1,..., q)$. This natural logarithm of odds ratio-based probabilistic framework is well-suited for analyzing accuracy measures, the advantage being that the framework uses all of the data (i.e., both successful and unsuccessful results in an experiment). Furthermore, the inverse logit function can be used to map fitted/predicted values on the log-odds scale back to the probability of interest, using the inverse of the logit function:

$$p(x) = \frac{e^{\alpha + \beta_1 x_1 + \beta_2 x_2 + \cdots + \beta_q x_q}}{1 + e^{\alpha + \beta_1 x_1 + \beta_2 x_2 + \cdots + \beta_q x_q}} \tag{3}$$

For more details on logistic regression modelling see Hosmer et al. [5]. What follows next is a description of the simulated dataset of experimental data, which is to be analyzed side-by-side for a comparison of repeated measures ANOVA and logistic regression modelling methods.

4   Description of the Dataset

In negative priming experiments of conceptual change, the leading memory-activating stimuli can be students' alternative conceptions or scientists' ideas about the same phenomenon. These fore-going primes' probabilistic influences on the following test trial items can be indexed by recording the participants' responses, especially correct proportions. An example of a study that could generate this type of data could be from international students' learning of the concept that the time of swing of a pendulum is the same (isochrony). However, in essence, the widely held notion of the isochrony of pendulum motion hinges on an easily ignored boundary condition: a small initial release angle ϴ. The reason is that only in that case the value of sin(ϴ) is roughly equal to ϴ itself. Out of such a boundary condition of a small ϴ, the pendulum motion is not isochronic anymore. Experimenters can design an experiment to test bilingual students' sensibility in detecting or ignoring such an anomaly.

Given the influence of the students' pre-instructional ideas, ignoring the anomaly of initial release angle ϴ and insisting on seeing the rod or string lengths as the only determining factor of a pendulum motion's isochrony tends to lead to a higher error rate and a faster choice response. Similarly, combining both the length and the initial release angle in such decision-making takes a little more time, which is expected to be recorded in the experiment for further analysis.

The dataset simulates a $2 \times 2 \times 2$ factorial design, where a prime target could be presented before or after a prime distractor ($T1_m$ vs. $T2_m$), and a probe-target could also occupy similar positions in probe displays ($T1_b$ vs. $T2_b$). The prime-to-probe repetition relationship could also vary between the ignored repetition condition (IR) and the unrelated repetition (UR). Table 1 lists some examples of these conditions. The IR condition is established to mimic the learning scenario where a scientific notion (a combination of an initial release angle ϴ and the length) is presented to a learner, but they keep using their alternative conceptions (only length) of the same pendulum motion phenomenon. This design is summarized in Table 1, representing a negative priming experiment for illustration.

Normally, it is expected that processing an unattended item (IR) would take longer than processing unrelated items (UR) because of distractor inhibition. The processing

Table 1 Example items from a simulated negative priming experiment

Prime couplet		Probe couplet		T1 position	T2 position	Prime-to-probe repetition
3	4*	四**	7	1	1	IR
3	4	六	7	1	1	UR
四	3	4	7	2	1	IR
四	3	6	7	2	1	UR
3	4	7	四	1	2	IR
3	4	7	六	1	2	UR
四	3	7	4	2	2	IR
四	3	7	6	2	2	UR

* The Italic indicates red distractors. In real experiments, the prime and probe couplets are to be presented along with uppercase letters fillers in rapid serial visual presentation. T1 refers to prime target position while T2 refers to probe target position. IR indicates ignored repetition, whereas UR indicates unrelated repetition
** If running a cross-language study, researchers can introduce non-alphabetic numerals in the Experiment

Table 2 Mean proportions of correct responses (%) of the simulated experiment

	T1 position 1		T1 position 2	
	T2 position 1	T2 position 2	T2 position 1	T2 position 2
IR	93.66 (1.18)*	89.25 (1.40)	93.24 (1.04)	91.57 (1.51)
UR	95.31 (1.11)	90.11 (1.55)	93.79 (0.95)	92.26 (1.53)

* Standard errors are in parentheses

also tends to lead to error-prone behaviours in the IR condition. To generalize from sample means differences to population means differences in these occasions requires an assumption of the underlying sampling distribution. If a normal distribution is assumed, the ANOVA framework shows the main effect of the factors and their interaction through sample means, between-group variances and within-group variances. Therefore, averaging cross-subjects and conditions is the first step in characterizing any dataset. Table 2 summarizes the mean identification accuracies and standard errors across the experimental conditions. However, does the result even make sense given that the responses can be viewed as either a question is answered correctly or the count of correct responses is averaged across the set of experimental conditions to represent the group mean levels? It does not. Moreover, the missing data and incorrect responses are often dropped during data preparation; thus, only correct responses are included for a group means comparison. All these routines are not always adequately justified for such data analyses. Consequently, a different modelling framework must be employed.

If a binomial sampling distribution is assumed instead, the same data could be analyzed through a logistic regression approach. As described above, logistic regression assumes a binomial distribution for the response. It uses a logit link function to directly model the probability of interest, permitting it to be expressed as a function of the other observed covariates/predictors, including interaction effects, on the log-odds scale. For comparison, the simulated dataset is analyzed first by applying the classic repeated-measures ANOVA (which only includes the correct responses and is deemed inappropriate to use in this context) and then using a mixed logit model (both the scientific concept level and participant level).

5 Analyzing Accuracy Data

5.1 Repeated-Measures ANOVA Over Untransformed Proportions Correct

A closer look at Table 2 reveals that the percentages correct of the simulated data are closer to 100% because a small set of familiar stimuli is used, which implies

Table 3 Summary of the repeated-measures ANOVA results over untransformed proportions correct (%) of the simulated experiment

Factors	$F(1, 21)$	p-value
T1 position (T1)	0.90	0.35
T2 position (T2)	10.03	0.005
Prime-to-probe relationship (PPR)	2.14	0.16
T1 × T2	8.15	0.009
T1 × PPR	0.32	0.58
T2 × PPR	0.20	0.66
T1 × T2 × PPR	0.14	0.71

the near-ceiling behaviours. From the perspective of sampling a binomial distribution, probabilities of these nearing-ceiling performances would behave differently than those of the responses clustered around the chance performance level (0.5). Statistically, it means that sample proportions close to fifty percent tend to be more homogeneous than those at the two far ends, in which one of them would represent the near-ceiling performance. Unequal variances of near-ceiling performances would violate the homogenous assumption of the repeated measures ANOVA. In addition, another look suggests that comparable correct responses are observed between the experimental and control conditions. It hints that it may be hard to detect any main effect of the experimental factors. Furthermore, it may not be easy to identify any significant interaction involving this factor.

As expected, a $2 \times 2 \times 2$ repeated-measures ANOVA over the untransformed accuracy has only revealed two significant extraneous effects: T2 position and T1 position × T2 position interaction. There are no other main effects or interactions revealed through applying repeated measures ANOVAs. Including the interaction effect is more informative than the main effect of T2 position. Processing an ignored item was more accurate when a response was made to a probe at T2 position. Conversely, more correct choices of an attended one were recorded when the probe was presented at T1 position. In other words, there is no simple position main effect since it also depends on the participants' attentional level.

Table 3 summarizes these results over untransformed accuracy data. As noted, the within-participants design of the conceptual change study has made the students' responses not necessarily independent anymore. It implies that ANOVA or a simple linear regression is no longer applicable. Would a logit model of the same data set also reveal a similar pattern when the analysis refocuses on the odds ratio of students' performance in the experiment?

5.2 Logistic Regression-Based Modelling of Accuracy Data

In contrast to the repeated-measures ANOVA, a logistic regression analysis is more flexible in modelling item-level and participant-level data with a mixed-effects modelling approach. In the case of conceptual change, the log odds of the probability of a participant's correct response can be conceptualized as a linear combination of the scientific conceptual item- and participant-related factors. In such a conceptualization, some aspects or parameters are the same for all participants (fixed effects), whereas the others are participant-specific (random effects). For example, having mastered the isochrony of pendulum motion, the students would increase their accurate responses. However, it depends on their information co-activating abilities, which differ for each. In combining the fixed and random effects together, a mixed effects model of conceptual change can be constructed to guide another data analysis, in which (1) a correct response is related to the factors and covariates by a specified link function, (2) non-normal distributions can be assumed, and (3) the repeated measures or correlated observations can be processed meaningfully.

The same dataset is also analyzed using a generalized linear mixed effect model by choosing a binomial probability distribution plus a logit link function. In other words, a mixed-effects logistic regression combines elements of logistic regression and linear mixed-effects models. In this way, the conceptual item-level and participant-level variability of this study can be considered simultaneously in the same model. More importantly, refocusing on odds rather than the frequentist's probability, the tendency to make incorrect responses can also be included within this approach to analyzing conceptual change results. In this sense, the analysis is carried out with the whole set of responses.

According to Bates et al. [1], the *lme4* R package offers the needed functions to fit linear mixed models with binary responses, assuming a model with both fixed- and random-effects terms. The fixed effects represent the effects of covariates one wishes to quantify. The random effects can account for the lack of independence in observations, such as repeated measures in the simulated data. With the R's *lme4* notation, the package can be installed by the R command install.packages("lme4") and then loaded into R using the command library(lme4). Now, the R functions such as lmer() for fitting linear mixed effects models (e.g., a mixed effects ANOVA) and glmer() for fitting generalized linear mixed effects models (e.g., a mixed effects logistic regression) are available for use. For example, the latter can be specified using the following command:

```
glmer(formula, data = NULL, family = gaussian,

control = glmerControl(), start = NULL, verbose = 0L,

nAGQ = 1L, subset, weights, na.action, offset,

contrasts = NULL, mustart, etastart, devFunOnly =

FALSE)
```

For this brief introduction, we can only focus on a few arguments of the glmer() function as described by Bates et al. [1]. The above formula describes both the fixed-effects and random-effects part of a model, with the response (*y* on the left, a ~

operator, and the regression terms associated with covariates (linearly connected by + operators on the right. A random-effects term can be specified using a vertical bar ("|" prior to the grouping factor. The term family specifies the underlying probability distribution, such as Gaussian (i.e., the normal distribution or binomial. Next, control is used to specify control structures for mixed effects model fitting, including specifying what optimizer to use and how to use it (for more details, see the R help files for lmerControl(or glmerControl(. At last, nAGQ specifies the number of points per axis for evaluating an approximation to the log-likelihood, which is used to fit the model (a value of 1 is the default and corresponds to a Laplace approximation; values greater than 1 will more accurately estimate the likelihood but take longer for the model to fit; and, a value of 0 uses a faster but less accurate method; see Bates et al. [1]. Together, they help express a mixed-effects logistic regression model. One of the possible such specifications can be given by the following command, which fits a generalized logistic mixed effects model to the cp data with fixed effects of position and PPR along with a random effect term for each participant to represent the repeated measures on each individual:

```
cc <- glmer(y ~ position + PPR + (1 | participant),

    data = cp, family = binomial,

    control = glmerControl(optimizer = "bobyqa"),

    nAGQ = 1)
```

Executing the command summary(cc) produces a summary of the resulting model fit, as illustrated in Table 4. The first section restates how the estimation was conducted. Here, I have collected these R steps in a single block for easy reading, with a model summary and input–output R libraries included also (see Appendix 1 for more details). Be aware that your computer file path specification may be different from what appears below.

```r
# Load required packages (these may need to be installed)
library(readr)# For importing data
library(lme4)# For GLMM modelling
library(jtools)# For helpful model summaries
library(rempsyc)# For exporting results


# Importing a simulated data set
cc_Raw <- read_csv("C:\\Data\\Data_Science.csv")
head(cc_Raw)


# A logistic GLMM-modeling of the simulated data set
cc1 <- glmer(response ~ position + ppr + (1|id),
    data = cc_Raw, family = binomial,
    control = glmerControl(optimizer = "bobyqa"),
    nAGQ = 1)
summary(cc1)
summ(cc1, exp=T)


# A second GLMM-modeling of the simulated data set
cc2 <- glmer(response ~ position + ppr + (1 + ppr|id),
    data = cc_Raw, family = binomial (link="logit"),
    control = glmerControl(optimizer = "bobyqa"),
    nAGQ = 1)
summary(cc2)
summ(cc2, exp=T)


# Nested Models Comparison
anova(cc1,cc2,test="Chisq")
```

Table 4 Results of model comparison results of the two embedded mixed effects models of the same simulated data

Model	AIC	BIC	LogLik	Deviance	Chisq	Df	Pr(>Chisq)
cp1	1205.2	1226.5	−598.6	1197.2			
cp2	1201.8	1233.7	−594.9	1189.8	7.3984	2	0.02474*

* $p < 0.05$

Note cp1 means a model built with R command "response ~ position + ppr + (1 | id)" whereas cp2 "response ~ position + ppr + (1 + ppr | id)."

The summary(cc1) output indicate evidence that the predictor position positively affects a student's correct responses, while the prime-to-probe relationship influences the responses differently. Specifically, compared to being a probe presented before a distractor, being presented at the other locations is more likely to select a correct choice. The prime-to-probe transition makes it less likely to select the correct one. Reading the value of such a model's parameter estimates requires exponentiating the estimates. The R library Jtools offers an easy way to do so through the command summ(cc1, exp=T). One notable aspect of interpreting these parameter estimates is that it is associated with the odds rather than probabilities.

In this example, two models have been built for comparison, with the first one containing only (1|id), whereas the second modelling an embedded and individualized prime-to-probe relationship (1 + ppr|id). The results of these two models can be compared with the anova command as illustrated in the above code snippets. AIC and BIC give us some information to tell these two models apart, with a smaller AIC indicating a better model fit. The results of model-comparison (see Table 4) show that the second embedded model has a smaller AIC and that there is moderate evidence ($0.01 < p$-value < 0.05) of improvement in fit compared to the first model, thus indicating a better mixed-effect logistic model.

For the same accuracy dataset, educational researchers have several options for analysis. Using ANOVA-based techniques helps reveal the main effects and their interactions, given the pre-defined experimental design. However, such a technique relies on the assumption that the response variable of interest is normally distributed. This is not the case when the response variable tracks correct/incorrect answers. For such accuracy data in conceptual change studies, logistic regression modelling provides the desired flexibility. In addition to identifying the same main effects, the repeated-measures aspect of the experiment can be modelled with the added flexibility of a logistic generalized linear mixed-effect regression model. Through the embedded model comparison, this logistic regression modelling approach helps validate that an embedded structure explains more variability than a simpler modelling framework. Combining the embedded model comparison and a logistic link function makes this approach to modelling accuracy data more informative and reliable.

Note that since the purpose of this chapter is not on the interactive phenomenon itself, the interaction of interest was not decomposed further into its componential contrasts. Nevertheless, it is evident that the logistic regression-based model can identify unique responding patterns that an ANOVA-based analysis suggested. Given that binary dependent variables such as correct or incorrect responses are readily available in conceptual change studies and that an S-shaped relationship between a learnt conception and the probability of correctly answering a test item might better capture the underlying conceptual change-induced responses, the generalized mixed effect linear models such as the logit modelling of accuracy measures should be preferred—and employed!—over ANOVA-type analyses in this context.

6 Discussion

The detectable structures in differences embedded in these well-performed responses may be tiny. Thus, modelling them using an ANOVA can easily distort response differences so that the actual effect may be subsequently masked out. Moreover—and more importantly—assuming a normal error distribution for binomial response data are inappropriate; this mismatch is more pronounced for analyzing the near-ceiling performance. Here, I have illustrated that the logit model may offer an appropriate alternative in this context because it considers the innate characteristics of dichotomous data and is sensitive to identifying accuracy differences from near-ceiling performances in conceptual change studies.

This tutorial has illustrated that the inappropriateness of ANOVA-based methods in conceptual change studies and has advocated for increased use of logistic regression-based modelling of accuracy data. Note, however, that dichotomous data are widespread in other contexts. They are expected in clinical research, such as attention deficit hyperactivity disorder (ADHD) or schizophrenia [7, 8]. Other analyses can involve responses such as choices, preference levels, or other categorical-coded variables. In all such situations, it is important to ensure that the modelling methods used to analyse such data are appropriate and that the assumptions underlying such methods are met; otherwise inferences and conclusions may be called into question.

7 Summary

The binary data generated from correct or incorrect responses should never be analyzed, assuming a normally distributed sampling distribution. Any inferences, including conclusions based on tests of significance, may not be correct due to the assumptions being violated. By showing the differences in analyzing accuracy data in conceptual change, we have demonstrated the advantages of applying a logit model for studying dichotomous outcomes. The benefit includes raw data modelling, fewer averaging artifacts, and tolerance of missing values. With abundant logit modelling techniques readily available, researchers do not have to keep relying on the blind use of variance analysis to explore accuracy-related data.

References

1. Bates, D., Mächler, M., Bolker, B., Walker, S.: Fitting linear mixed-effects models using lme4. J. Stat. Softw. **67**, 1–48 (2015). https://doi.org/10.18637/jss.v067.i01
2. Chi, M.T.H.: From things to processes: a theory of conceptual change for learning science concepts. Learn. Instr. **4**, 27–43 (1994)
3. diSessa: Toward an epistemology of physics. Cognit. Instruct. **10**(2/3), 105–225 (1993)
4. Golay, P., Fagot, D., Lecerf, T.: Against coefficient of variation for estimation of intraindividual variability with accuracy measures. Tutorials Quant. Methods Psychol. **9**(1), 6–14 (2013)
5. Hosmer, D.W., Lemeshow, S., Sturdivant, R.X.: Applied Logistic Regression, 3rd edn. Wiley (2013)
6. Jaeger, T.F.: Categorical data analysis: Away from ANOVAs (transformation or not) and towards logit mixed models. J. Mem. Lang. **59**(4), 434–446 (2008)
7. Kalff, A.C., De Sonneville, L.M., Hurks, P.P., Hendriksen, J.G., Kroes, M., Feron, F.J., Steyaert, J., Van Zeben, T.M., Vles, J.S., Jolles, J.: Speed, speed variability, and accuracy of information processing in 5 to 6-year-old children at risk of ADHD. J. Int. Neuropsychol. Soc. **11**(02), 173–183 (2005)
8. Knouse, L.E., Bagwell, C.L., Barkley, R.A., Murphy, K.R.: Accuracy of self-evaluation in adults with ADHD evidence from a driving study. J. Attent. Disorders **8**(4), 221–234 (2005)
9. Kuffner, T.A., Walker, S.G.: Why are p-values controversial? Am. Stat. **73**(1), 1–3 (2019). https://doi.org/10.1080/00031305.2016.1277161
10. Potvin, P., Sauriol, É., Riopel, M.: Experimental evidence of the superiority of the prevalence model of conceptual change over the classical models and repetition. J Res Sci Teach. **52**(8), 1082–1108 (2015). https://doi.org/10.1002/tea.21235
11. Schneider, M., Vamvakoussi, X., Van Dooren, W.: Conceptual change. In: Seel, N.M. (ed.) Encyclopedia of the Sciences of Learning, pp. 735–738. Springer US (2012). https://doi.org/10.1007/978-1-4419-1428-6_352
12. Thagard, P.: Concepts and conceptual change. Synthese **82**(2), 255–274 (1990). https://doi.org/10.1007/BF00413664
13. Vosniadou, S.: Capturing and modeling the process of conceptual change. Learn. Instr. **4**(1), 45–69 (1994). https://doi.org/10.1016/0959-4752(94)90018-3
14. Wasserstein, R.L., Lazar, N.A.: The ASA statement on p-values: context, process, and purpose. Am. Stat. **70**(2), 129–133 (2016). https://doi.org/10.1080/00031305.2016.1154108

Transforming Data on the Boundaries of Science and Policy: The Council of Canadian Academies' Rhetorical Repertoire

Matthew A. Falconer

Abstract The following qualitative case study presents ways that a boundary orga- nization—the Council of Canadian Academies—works between the domains of science and policy to transform and re-purpose data for government policy-makers. How such an organization manages to do this remains obscured. The chapter features an analysis of three reports produced by the CCA in 2017, 2019, and 2021 for three separate Canadian governmental bodies, and explores how data are taken-up from scientific domains and transformed for policy-makers. Findings reveal the rhetorical repertoire employed by a boundary organization in transforming data.

Keywords Recontextualization · Boundary work · Rhetorical strategies · Data

1 Introduction

As the world faces increasingly complex issues, the need to incorporate data into decision-making grows each day [10, 18, 22]. There has been a considerable atten- tion given to the work of scientists in producing data-driven knowledge [23, 39]. A common view of the rhetorical nature of science is that the research lifecycle constructs scientific knowledge [19, 30]. Data translators—those who assist with transforming and re-purposing or "discursively recontextualizing" [26] numeric information—play an important role in brokering data. Such work depends on the uptake [14, 15] of data by actors in a separate domain, a type of work often done by "knowledge brokers" [29].

Social actors working on the "boundaries" [16, 17] of science often perform such brokering [2, 25]. This is known as "boundary work", which is a discursive activity aimed at bridging the communicative gap between the conceptual worlds of science and policy [21]. Boundary work is often carried out by organizations working between boundaries, or "boundary organizations" [21]. One such boundary organization is the Council of Canadian Academies (the CCA). Based in Ottawa,

M. A. Falconer (✉)
School of Linguistics and Language Studies, Carleton University, Ottawa, ON, Canada
e-mail: matthewfalconer@cunet.carleton.ca

Canada, the CCA uses a multidisciplinary and multi-sectoral expert panel model to produce reports for clients, which are typically federal governmental departments. Such boundary work involves discursively recontextualizing [26]—that is, transforming and repurposing—science and other types of evidence for its clients, which are typically government policy-makers [13]. This boundary work leads to the creation of "boundary objects" [40, 41], which are often written products that cross such discursive boundaries and are used by actors in a separate domain for their own purposes.

More recently, researchers have explored the use of language to share scientific data from a number of perspectives, including: the uses of visuals [20, 31, 42], the translation of science [24], and the recontextualization of science or the ways that scientific knowledge is transformed and re-purposed among expert scientists [27, 37, 38], to name but a few. However, little is known about how data are transformed and re-purposed for policy-makers by boundary organizations [13]. This chapter contributes to this end by exploring how one boundary organization—the CCA—transforms and re-purposes data for government policy-makers.

The study responds to the following questions: In what ways is data transformed and re-purposed by the CCA? What do CCA clients receive? What rhetorical strategies are employed by the CCA when transforming and re-packaging data for its clients?

In responding to these questions, I suggest that the CCA has an identifiable repertoire of data translation strategies they use to cross disciplinary boundaries and offer policy-relevant information for government policy-makers. Throughout the chapter, I use the terms transformation and re-purposing—recontextualizing—when describing the CCA's data translation work. The tools and approaches the CCA uses and their rhetorical repertoire offers insights into real-world examples of data translation in-action. I offer observations on the rhetorical strategies that appear in three reports—each being boundary objects produced by the CCA—as ways the organization has developed to recontextualize relevant data for its clients. I present my argument first by identifying relevant research and theories that inform the chapter. Then, I present my methodology and the findings of the study.

2 Related Research and Theories

For my discussion, I survey related research and theories from the rhetoric of science, and the ways that knowledge is transformed between expert and non-expert domains. The research and theories generally fit within a social constructionist framework [3].

2.1 Rhetoric of Science

The rhetoric of science is a field of study that explores how language is used to successfully accomplish goals within the domain of science [1, 39]. A contemporary view of rhetoric considers the ways people use language to achieve any particular goal [1, 4]. Importantly, these views have seen language as involving any symbol system used to convey a message to a particular audience, including numbers and visuals [37]. Rhetoricians of science—those who study language use in science—have developed insights into the nature of how language is used within and outside of science using these insights [39]. In this area, rhetoricians have explored different strategies that scientists employ in communicating their research [1, 12, 30]. In this view of rhetoric, all things that are included in a scientific communicative act—including data—are open for investigation.

Within rhetoric, different scholars have looked to the different strategies that people use when trying to successfully convey their message [11, 13, 38]. Such "rhetorical strategies" are the ways that people communicating a message attempt to persuade the audience in different settings. Earlier, [13] explored the rhetorical strategies used by the Council of Canadian Academies in one report [5] when presenting data. He found three rhetorical strategies surrounding the use of visual representations to characterize people, synthesizing and depicting trends over-time from one or more data sets, and written explanations of data. In the following discussion, I use this as a starting-point for my analysis and present findings that extend these insights.

2.2 Data Translation as a Semiotic Process of Uptake and Recontextualization

The recontextualization of knowledge involves transforming and re-purposing pre-existing ideas and information into a new form of knowledge that serves a different purpose [8, 26, 35]. This is different than data translation, which aims to make accessible data to various audiences. Those who have employed this understanding of recontextualization have explored how scientific knowledge is transformed and re-purposed in several domains, including scientific, engineering, healthcare, and legal contexts [27, 32, 36]. Others have explored the way that science is made accessible as it is recontextualized for non-expert audiences [28, 33].

More recently, [38] suggested that recontextualization is a single, two-step semiotic process involving "uptake" and recontextualization. In their view, a social actor conveys an idea first in a specific text within their own context. This text is then made available to other social actors. Then, a separate social actor "takes-up" [14] an idea from a pre-existing source of information or knowledge. Once they take the ideas into their own discursive work, a social actor then transforms and re-purposes the knowledge. In doing so, they create new meanings that in-turn inform or influence others in their discursive activity, ways of thinking, and/or related human activities

[15]. Later, I use these notions to examine the types of data the CCA takes-up in an Expert Panel Report and consider the ways that data are transformed and re-purposed.

3 Methodology

This chapter presents a qualitative case study [9] of the Council of Canadian Academies (the CCA) data translation work. I focus on the use of data sources in a corpus of texts I gathered and analyzed in-depth—a textual analysis [39]—that the CCA produced for different government departments. Here, I present an overview of the data collected and analytical approaches used.

3.1 Data Collected

To explore the ways that the CCA transforms data for government policy-makers, I collected three "Expert Panel Reports". These reports include:

- *Older Canadians on the Move* [5], which explores ways of adapting the Canadian transportation system to meet the needs of an aging population;
- *When Antibiotics Fail* [6], which describes the potential socio-economic impacts of antimicrobial resistance in Canada; and,
- *Degrees of Success* [7], which discusses the labour market transition of PhD graduates in Canada.

These three reports were gathered online for free from the CCA's website (https://cca-reports.ca/) in summer 2021. These reports were given to three different federal governmental departments and agencies—Transport Canada [5], The Public Health Agency of Canada [6], and Innovation, Science and Economic Development Canada [7] respectively—at two-year intervals.

After surveying all available CCA reports published from 2006 through 2021 (the time of writing), I chose these three reports for the breadth of subject areas reflected in each document. The reports cover various disciplinary focuses ranging from biochemistry, economics, education, engineering, epidemiology, health science, microbiology, medicine, sociology, among others. Such breadth allows for a discussion sensitive to how data taken from several different disciplinary domains are transformed and re-purposed by a boundary organization working between the domains of science and policy.

Readers unfamiliar with qualitative research may find the sampling approach that I used to be biased. To some extent, that is a fair point, at least from a quantitative perspective. Qualitative research of this nature often relies on the researcher's selection. For readers coming from quantitative paradigm, take the findings as an explorative study of how scientific data are transformed for policy-makers. These findings are observational, and rely on interpretive frameworks from well-established disciplinary concepts presented above. Further research could explore more robust sampling techniques.

3.2 Analytical Approaches

For my textual analysis of how data are transformed by the CCA, I extend an earlier line of thinking of the rhetorical strategies—that is, the purpose for which the data are re-purposed in the report—used in one report, *Older Canadians on the Move* [13]. Earlier, I found three strategies used to transform data sources in *Older Canadians on the Move*, which are listed above. I used this as a starting point and applied the lens to the other Expert Panel Reports considered here. Specifically, I extend my analysis of *Older Canadians on the Move* to consider the uses of data sources in all three reports [5–7]. I use the larger sample size to extend my earlier understanding of recurring rhetorical strategies that the CCA uses to transform data.

To establish these categories, I adapted qualitative coding techniques [34] to systematically categorize passages of the reports focusing specifically on excerpts featuring discussions of data. I first used a deductive coding approach for my textual analysis using pre-identified categories of evidence. I began with my earlier work that identified culture-specific types of evidence—academic literature, grey literature, external data sources, and media sources—used by the CCA [13].

For this study, I focused specifically on the uses of external data sources, which is one type of evidence that the CCA views as credible. These data sources are typically not produced by the CCA, rather they are often publicly available data sets from reputable organizations, like Statistics Canada or the Organization for Economic Co-operation and Development.

To focus on external data sources, I followed the citation trail in each report. I examined the citations and reviewed the source to determine that it was an external data source. Once I identified the external data sources, I then explored rhetorical strategies that the CCA used. To do so, I did further deductive coding using pre-identified rhetorical strategies [13]. With a larger sample of texts, I also did inductive coding to challenge and expand earlier findings. These strategies are featured in the findings section below.

4 Findings

In this section, I present the findings of my study of the Council of Canadian Academies' (the CCA). I begin with a discussion of what the CCA uses as sources of data. Then, I look at four of the CCA's rhetorical strategies for re-purposing data into a form that is useable by government policy-makers: representations of people, visualizations of data, written explanations, and numeric representations.

4.1 Data Uptake

In the CCA, data are one of four common sources of evidence used to respond to client-initiated questions. The other three common sources of evidence are peer-reviewed academic literature, grey literature, and media stories. The CCA typically uses quantitative data. However, as we will see below, there are instances where the client-initiated questions require qualitative data. The CCA does incorporate such data, and re-purposes it for the report.

Data typically is created and maintained by other researchers and organizations and is brought into the CCA's panel process [13]. The CCA values quality data sources and seeks publicly available data sets from reputable organizations. This data in its raw form comes as data tables. These are typically governmental organizations (e.g., Statistics Canada) and intergovernmental organizations (e.g., the Organization for Economic Co-operation and Development). All three reports included in this study used data sets from these two organizations.

For all data, each report features original analysis conducted during the expert panel process. In each case, the Expert Panel determines what methods to use, which creates room for variability. For example, *When Antibiotics Fail* designed its own analytical approach:

> ...the Panel also commissioned a quantitative economic model to explore the complex rela-
> tionship among AMR [anti-microbial resistance], health, labour productivity, agriculture,
> and trade (Sect. 4.1 and Appendix C). The model considers production and trading patterns
> among industries and countries, assuming that if AMR affects labour productivity in one
> country, global production and trade adjust throughout the world [6, p. 4].

In this example, we see the CCA adopting formal analytical methods employed by economists to create an original transformation and interpretation of existing data.

There are other types of data sources used by the CCA. The specific data used seems to depend entirely on the client-initiated question guiding the report writing process. For example, *Degrees of Success* draws on data and reports from provincial governments and individual Canadian universities to identify employment outcomes [7, pp. 61–62]. Also, *When Antibiotics Fail* uses data generated by health-related organizations, such as the Canadian Institute for Health Information, the Public Health Agency of Canada, and the World Health Organization.

4.2 Recontextualizations of Data

The types of data described above are "taken-up" [14] into the CCA's discursive activity of producing an Expert Panel Report [13]. What we have access to is the final product—the reports themselves—that function as a type of "boundary object" [41], or an artefact that crosses the science-policy boundary and is useable in either domain without becoming something entirely new. Data are recontextualized in those reports. Below, I discuss aspects of the CCA's rhetorical repertoire. These include representations of people, visualizations of data, written explanations, and numeric representations. For each, I also identify the rhetorical strategies employed as data are recontextualized for government policy-makers.

4.2.1 Representations of People

One of the ways that the CCA recontextualizes data are to use visual call-out boxes to represent specific parts of the population. This rhetorical strategy appears in two of the reports to depict parts of a population and summarize quantitative data [5, 6], and to highlight individual experiences and feature qualitative data [7]. These call-out boxes seem to draw the reader's attention to the specific example as each features stark colours on the page.

In *Older Canadians on the Move*, the CCA drew on a type of marketing research method to create "personas". These personas use stories to depict and summarize certain commonalities within a part of a population. They are used to summarize four types of older Canadians travelling in Canada and use narrative to explain the situation. Below is an example of one of the four personas used in that report (Fig. 1).

When Antibiotics Fail features a similar strategy through "vignettes" in Chap. 5. These vignettes serve a similar purpose to the personas found in *Older Canadians on the Move* in generalizing specific parts of the population and their needs in relation to antimicrobial resistance. Figure 2 below demonstrates this:

Where both other reports feature generalized depictions of issues people within the population face, *Degrees of Success* features "personal narratives" using qualitative

Fig. 1 A persona for "Yumi" as depicted in Older Canadians on the Move (reprinted from CCA with permission)

> **YUMI (73 years old)**
> *British Columbia*
>
> Yumi is a divorced Japanese-Canadian who immigrated to British Columbia in 1972 with her former husband. After Yumi's divorce, money grew tight; because she was a homemaker while raising her daughter, Misato, the only work she could find after the divorce was a minimum-wage cashier job. Yumi now lives alone in a small apartment in Burnaby.
>
>
>
> Misato lives in Seattle with her husband and son. Now that their son is older, Misato and her family have many weekend commitments and rarely get a chance to visit Yumi. Yumi misses them and, having two weeks' worth of unused vacation, she considers travelling down to Seattle by train for the first time.
>
> Yumi has a number of concerns about the trip:
> - How will she manage planning the trip and travelling alone? She has never been to the train station and would have to travel by bus to get there.
> - Can she afford the train ticket? Will she be able to bring food on the train to avoid having to buy an expensive meal?
> - How will she contact Misato when she arrives? Will the prepaid cell phone Misato gave her work outside Canada?

> **Vignette: Renée**
>
> Renée is a 52-year-old from Winnipeg who has
> been suffering from osteoarthritis for many
> years. She was forced to retire early from her
> job as a nurse when her symptoms became
> too debilitating. Due to decreased mobility,
> Renée developed obesity and type 2 diabetes.
> Her left hip is the most severely affected joint
> and her doctor says that, in the past, a total hip
> replacement would have been the recommended
>
>
> Stock photo. Posed by model.
>
> course of action. However, the doctor is reluctant to suggest surgery as resistance
> continued on next page

Fig. 2 Part of a vignette from When Antibiotics Fail (reprinted from CCA with permission)

Fig. 3 An example of a
personal narrative from
Degrees of Success
(reprinted from CCA with
permission)

PHD PATHWAY 6
Lisa Bélanger, CEO, ConsciousWorks (2017)

Lisa Bélanger obtained a PhD in behavioural medicine from UofA.
During her doctoral work, she became passionate about how "seemingly
small behaviours" could have huge impacts on human health. Bélanger
decided that, after graduation, she wanted to "create impact" around
the research she carried out at UofA. Her first job after graduation
was at a research funding agency, but the position was not a good fit.
Still, she felt she left the position with "valuable connections and skills
development."

Bélanger is now CEO of ConsciousWorks, a consulting company she
founded that "teaches people how insights from behavioural science
can improve their personal and professional lives." She also founded
a charity called Knight's Cabin, which provides "no cost, research-
based retreats focused on physical activity, nutrition, stress, and sleep
for cancer survivors and their supporters across Canada." Bélanger
continues to sit on the board of directors of Knight's Cabin as founder
and scientific director. Entrepreneurship was not new to Bélanger, who
started a small personal training company during graduate school and
enjoyed the challenge and flexibility of working for herself.

Bélanger notes that the thing she has found most surprising in her career
is the importance of networks, both for her company and her charity.
"Your network is your first, and most valuable, asset," Bélanger explains.

Adapted from Polk (2017c)

data. These narratives are text-based and describe real-world experiences of 11 people
"…to convey the individual experiences of PhD students and graduates from a variety
of disciplines" [7, p. 6]. Figure 3 provides an example of one such narrative.

4.2.2   Visualizations of Data

The CCA uses at least two strategies for visually representing data. First, the CCA
regularly uses graphs to represent trends over-time based on one or more data sets.

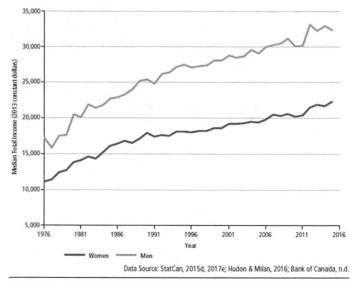

Figure 3.3

Median Total Income of Adults Aged 65 or Older in Canada, Divided by Sex
The median income (in 2013 constant dollars) of men (blue line) and women (orange line) aged 65 or older in Canada from 1976 to 2015. Data for 2014 and 2015 were converted to 2013 constant dollars using the Bank of Canada Inflation Calculator in September 2017.

Fig. 4 An example of a visualization of data found in Older Canadians on the Move (reprinted from CCA with permission)

Though there are many examples of other types of graphs throughout each report (e.g., bar and pie graphs), line graphs show multiple data points that indicate rising trends overtime (Figs. 4 and 5 below).

Another approach to visually representing data are an infographic. In the examples below (Figs. 6 and 7), the data are displayed to depict specific numeric information. Together, the visualizations depict a complicated flow of statistics that breaks down each bit of information in a manner that is quick for policy-makers to understand. Figure 6 uses pie graphs, generic figures resembling people, and large fonts to deliver the main message. Figure 7 uses generic people with percentages in clear, colourful bubbles beneath to depict segments of the population with specific needs.

4.2.3 Written Explanations

Another rhetorical strategy used by the CCA in recontextualizing data involves representing and describing information in written discourse. This strategy appears constantly in all of the reports considered here, especially in sections that provide a written description of predominantly statistical information. Here are three examples of this strategy, one taken from each report:

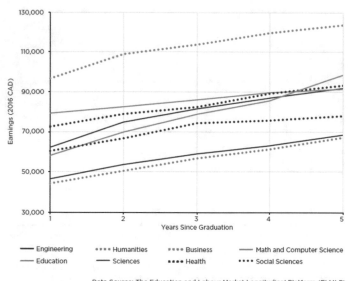

Data Source: The Education and Labour Market Longitudinal Platform (ELMLP)

**Figure 5.2 Earnings Trajectory of PhD Graduates by Field of Study,
2010 Cohort**

Fig. 5 Example of data visualization in Degrees of Success (reprinted from CCA with permission)

1. The population age structure among provinces and territories is highly variable.
   The percentage of the population over 65 is higher in the Atlantic provinces,
   British Columbia, and Quebec compared to the Canadian average, while it is
   lower in Alberta and the territories (StatCan, 2017b). At the extremes, as of
   2016, 20% of Nova Scotians were over 65 while the same was true for only 3.8%
   of Nunavummiut (StatCan, 2017b). The proportion of Indigenous adults aged 65
   or older was about 6% in 2011, compared to over 14% for non-Indigenous adults
   in the same year (StatCan, 2013) [5, p. 26].

2. The estimates in Table 2.3 demonstrate that AMR is already a serious problem
   in Canada and responsible for significant negative health outcomes. Based on
   the 10 important clinical syndromes, people in Canada acquired at least 250,000
   resistant infections in 2018; 7 over 14,000 deaths were caused by infections
   that were resistant to first-line treatment (Fig. 2.4). Furthermore, 5,400 of these
   deaths, or almost 15 a day, could be considered to be directly attributable to
   AMR itself. In other words, these 5,400 deaths would not have occurred had the
   people had a susceptible infection. These results suggest that 4 out of every 10
   deaths from a resistant infection would not have occurred if the infection were
   not resistant (Fig. 2.4). In 2018, AMR was therefore the attributable cause of
   only slightly fewer deaths in Canada than Alzheimer disease in 2016 (StatCan,
   2019) [6, p. 34].

3. The number of people graduating with PhDs in Canada has been increasing at a
   relatively steady rate since 2002. In that year, 3,723 students graduated from PhD

Types of Infections in Canada 2018
In 2018, there were approximatively 980,000 bacterial infections in Canada
Resistant Infections
(n = 250,000)
Susceptible Infections
(n = 730,000)
Resistant Infections
Susceptible Infections
Of 250,000 resistant infections
14,000 people died
Of 730,000 susceptible infections
30,000 people died
Of these 14,000 deaths:
4 in 10 would not have occured if the infection was susceptible to first-line antimicrobials
Figure 2.4
Resistance and Mortality of Bacterial Infections in Canada for the 10 Important Clinical Syndromes
The Panel estimates that there were just under one million bacterial infections in Canada in 2018 leading to one of the 10 important syndromes. Approximately a quarter of these infections were resistant to first-line antimicrobials. In that same year, it is estimated that approximately 4 in 10 of those who died from resistant infections would have survived were their infection susceptible to first-line antimicrobials.

Fig. 6 An example of an infographic from When Antibiotics Fail (reprinted from CCA with permission)

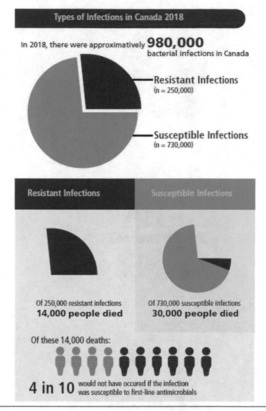

or equivalent programs across the country (StatCan, 2020 h). The number rose to 5,946 by 2010 and by 2017 had more than doubled, reaching 7,947. The number of PhD graduates has continued to grow at a rate faster than the population in Canada aged 17 to 64 (which encompasses the majority of people in the labour market) (StatCan, 2020a, 2020 h) (Fig. 2.1). The rate of growth for PhD graduates is also higher than the growth rate for bachelor's graduates. For instance, while there was a 113% increase in the number of PhD graduates between 2002 and 2017, the increase in bachelor's graduates was 52% (StatCan, 2020h) [7, p. 12].

DIFFICULTY LIFTING
MORE THAN 4.5 KG
7%   11%   22%

DIFFICULTY STANDING
FOR MORE THAN 15 MIN
16%   23%   36%

DIFFICULTY WALKING
2 TO 3 BLOCKS
11%   19%   32%

DIFFICULTY STANDING
UP AFTER SITTING
23%   28%   39%

DIFFICULTY REACHING
ARMS ABOVE SHOULDERS
11%   15%   25%

DIFFICULTY SITTING
FOR 1 HOUR
10%   9%   7%

Age: 65–74 YRS   Age: 75–84 YRS   Age: 85+ YRS

Data provided by the Canadian Longitudinal Study on Aging

Figure 3.1

Older Adults in Canada Who Have Difficulty Performing Certain Tasks

The percentage of older adults who have difficulty with or cannot perform specific tasks, stratified by age. Percentages are based on the number of respondents. Data was collected from 2011 to 2015.

Fig. 7 An example of an infographic from Older Canadians on the Move (reprinted from CCA with permission)

Table 3.2

Most Common Forms of Transportation Among Older Adults in Canada Who Do Not Drive

Age	In the past year, which was your most common form of transportation?				
	Passenger in a motor vehicle (%)	Public or accessible transit/ taxi (%)	Cycling/walking (%)	Wheelchair or scooter (%)	Do not know/no answer or refused (%)
65–74 (n=706)	38.39	42.92	14.59	1.98	2.12
75–84 (n=1,073)	50.70	35.88	9.41	2.33	1.68
85+ (n=87)	58.62	32.18	4.60	0.00	4.60

Data provided by the Canadian Longitudinal Study on Aging

This table shows the most common form of transit in the past year for participants who do not have a driver's licence (currently or never), who responded with a "Don't know/No answer" or refused to respond to the driving status question, or have a driver's licence but never drive, stratified by age. The number of respondents by age category is provided. The table reports percentages based on the number of respondents. Data collected by the CLSA from 2013 to 2016.

Fig. 8 An example of numeric representations in Older Canadians on the Move (reprinted from CCA with permission)

4.2.4 Numeric Representations

The CCA also employs numeric representations in the form of tables. This rhetorical strategy summarizes relevant data related to the report's argument in a succinct manner. Most often, the numeric information are descriptive statistics (Figs. 8, 9). However, there are times when the CCA presents raw numbers as numeric representations as well (Fig. 9). This strategy was found in each of the three reports, as the Figs. 8, 9, and 10.

4.3 Brief Discussion of Findings

These findings reveal the different ways that the CCA takes-up data from external sources and recontextualizes the information in a manner unique to the report— a boundary object—for government clients. The analysis suggests that there is a common set of criteria that the CCA uses to determine whether a data source is credible and warrants being taken-up. There are recurring data sources the CCA seeks out, and specific types of data that is needed for different projects. These criteria are flexible so long as the data comes from a reputable source [13]. These data sources are defined by the boundary work performed by the CCA—the data are reputable to both scientific researchers and government policy-makers.

It also reveals the ways that data are transformed and re-purposed in the final reports that the CCA delivers to government policy-makers. The findings suggest that there is a repertoire within the CCA of how to recontextualize data. There are specific types of science-based tools used to re-purpose data. These include tools for representing people, visualizing data, writing about data, and providing numeric representations. These tools re-purpose data in a way that fits a report's narrative. These tools are used in each report, regardless of the department who requested the expert panel be convened. Taken together, they suggest a series of approaches that the

Table 4.3

Canadian Estimates of Lost Years of Employment, 2018

Syndrome	Total Deaths	Attributable		
		Deaths	LOS (Work Years)	Lost Employment (Work Years)
BGI	3	1	28	29
BSI	630	221	134	355
CDI	211	115	255	370
IAI	2,266	1,956	439	2,395
MSI	376	231	268	499
Pneumonia	2,072	366	108	474
SSTI	1,781	636	1,018	1,654
STI (gonorrhea)	1	1	4	5
TB	7	0	45	45
UTI	6,732	1,900	1,216	3,116
Total	14,000	5,400	3,500	8,900

This table presents total deaths, attributable deaths, and attributable length of stay (LOS) in hospital for resistant infections (in work years), as estimated in Table 2.2. Lost years of employment are calculated as the number of attributable deaths plus LOS. For example, resistant UTIs are estimated to have caused 1,900 deaths and 1,216 years (304,045 days) in hospital in 2018, which translates to 3,116 lost years of employment (where one year equals 250 days in LOS). See Table 2.3 for references. BGI, bacterial gastrointestinal infection; BSI, bloodstream infection; CDI, *C. difficile* infection; IAI, intra-abdominal infection; MSI, musculoskeletal infection; SSTI, skin and soft tissue infection; STI, sexually transmitted infection; TB, tuberculosis; UTI, urinary tract infection. The totals for deaths and LOS are rounded due to the uncertainty in the estimates.

Fig. 9 An example of numeric information in a table from When Antibiotics Fail (reprinted from CCA with permission)

CCA has developed through trial and error to convincingly re-purpose data through its boundary work.

The various rhetorical strategies that were found in-use for each of these ways to re-purpose data have been identified above in each subsection. These different strategies make-up the CCA's rhetorical repertoire for recontextualizing data on the boundaries of science and policy. Each strategy has been used intentionally to re-purpose data in a way that makes complex information much more accessible to general readers. They remove the analytical process needed to understand rich data, and present the data in a manner that limits the amount of cognitive effort needed to get the meaning of otherwise complex information. The rhetorical strategies are representations of largely quantitative data from single or multiple sources that pull relevant information together in a succinct fashion.

Table 4.1   Percentage of Bachelor's, Master's, and PhD Graduates
Working in Education, by Gender and Census Year

	Men			Women		
	2006	2011	2016	2006	2011	2016
Bachelor's	7.5%	7.1%	5.9%	18%	17%	15%
Master's	14%	13%	11%	23%	23%	20%
PhD	40%	40%	39%	46%	44%	45%
PhD (under 40 only)	43%	40%	41%	45%	39%	41%

Data Source: Panel analysis of PUMF data for the 2006 and 2016 Canadian censuses, and the 2011
National Household Survey. Individuals are defined as working in education if they hold an education-
related position in an education industry or sector.

Fig. 10 Example from Degrees of Success (reprinted from CCA with permission)

5 Conclusion

This chapter has revealed the ways that a boundary organization—the Council of
Canadian Academies (the CCA)—transforms and re-purposes data sources taken
from various places and used in the reports they produce for government policy-
makers. Specifically, it offers insights into the rhetorical strategies found in reports
produced by an organization that relies on interdisciplinary expertise to complete
its work. In looking at three reports [5–7], the study offers insights into the rhetorical repertoire of a boundary organization that transforms data into a distinctly new
form of written knowledge, where new meanings are developed. It also provides
additional support for the view of recontextualization as being a single, two-step
semiotic process—that is, a meaning-making process involving language use—that
involves taking-up and transforming prior knowledge into a new set of meanings
[38].

This chapter offers a Discourse/Writing Studies view to the current volume.
Specifically, it provides insights into the nature of interdisciplinary data translation for government policy-makers. In doing so, I have provided an account of the
rhetorical life of a boundary object that incorporates complex, diverse sources of data
considered relevant by the CCA, a boundary organization. I have identified some of
the ways that data are re-purposed in a way that meets the expectations of policy-
makers, which is apparent by the CCA's ongoing funding from the Canadian federal
government. Future research could address the perspectives of policy-makers who
receive reports like the CCA's, and how helpful they find the tools and rhetorical
strategies identified above.

Pedagogically, the chapter offers a series of real-world examples from one organization's boundary work between science and policy. By understanding the rhetorical repertoire of available strategies used by the CCA to transform and re-purpose

data, those in teaching and learning environments have examples of what such work looks like. Though observational, such a perspective offers insights into what those working in data analysis positions may expect to do when working in an interdisciplinary and inter-sectoral environment. The CCA is one boundary organization doing this work, but the type of data translation is consistent with best practices carried out by similar organizations. Seeing and understanding the discourse of data translation better positions readers to learn go-to approaches, and challenge established norms for data translation practices as new insights and practices continue to emerge.

Matthew Falconer is an Adjunct Research Professor in Carleton University's School of Linguistics and Language Studies. His research explores the uses of science in non-scientific domains. His doctoral research involved an ethnography of the Council of Canadian Academies' collaborative discursive boundary work of transforming science for policy-makers. He has published on the uses and misuses of science in the ongoing public climate change debate, and knowledge-making practices in science.

References

1. Bazerman, C.: Shaping Written Language: the Genre and Activity of the Experimental Article in Science. University of Wisconsin Press, Madison (1988)
2. Bednarek, A.T., Shouse, B., Hudson, C.G., Goldburg, R.: Science-policy intermediaries from a practitioner's perspective: the Lenfest Ocean program experience. Sci Public Policy 43, 291–300 (2016)
3. Berger, P.L., Luckmann, T.: The Social Construction of Reality: a Treatise in the Sociology of Knowledge. Penguin, New York, NY (1967)
4. Burke, K.: Language as Symbolic Action. Cambridge University Press, Cambridge, UK (1966)
5. CCA (Council of Canadian Academies).: Older Canadians on the move. Council of Canadian Academies, Ottawa, ON (2017)
6. CCA (Council of Canadian Academies).: When antibiotics fail. Council of Canadian Academies, Ottawa, ON (2019). https://cca-reports.ca/reports/the-potential-socio-economic-impacts-of-antimicrobial-resistance-in-canada/
7. CCA (Council of Canadian Academies).: Degrees of success. Council of Canadian Academies, Ottawa, ON (2021). https://cca-reports.ca/reports/the-labour-market-transition-of-phd-graduates/
8. Coupland, N., Coupland, J.: Reshaping lives: constitutive identity work in geriatric medical consultations. Text 18(2), 159–190 (1998)
9. Creswell, J.W., Poth, C.N.: Qualitative Inquiry and Research Design: Choosing Among Five Approaches, 4th edn. SAGE, Los Angelas, CA (2016)
10. Douglas, H.: Weighing complex evidence in a democratic society. Kennedy Inst. Ethics J. 22(2), 139–162 (2012)
11. Fahnestock, J.: Accommodating science the rhetorical life of scientific facts. Writ. Commun. 3(3), 275–296 (1986)
12. Fahnestock, J.: Rhetoric of science: enriching the discipline. Tech. Commun. Q. 14(3), 277–286 (2005)
13. Falconer, M.: Providing science advice: an ethnography of the Council of Canadian Academies' discursive boundary work of recontextualizing science for policymakers. Unpublished doctoral dissertation, Carleton University, Ottawa, ON (2019). https://curve.carleton.ca/cbcf2142-a876-4c8c-aa36-a9f74e869b8d

14. Freadman, A.: Anyone for tennis? In: Freedman, A., Medway, P. (eds.) Genre and the New Rhetoric, pp. 43–66. Taylor & Francis, London, UK (1994)

15. Freadman, A.: Uptake. In: Coe, R., Lindgard, L., Teslenko, T. (eds.) The Rhetoric and Ideology of Genre, pp. 39–53. Hampton, Cresskill, NJ (2002)

16. Gieryn, T.F.: Boundary-work and the demarcation of science from non-science: strains and interests in professional ideologies of scientists. Am. Sociol. Rev. **48**(6), 781–795 (1983)

17. Gieryn, T.F.: Boundaries of science. In: Jasanoff, S., Markle, G.E., Peterson, J.C., Pinch, T. (eds.) Handbook of Science and Technology Studies, pp. 393–443. Sage, Thousand Oaks, CA (1995)

18. Gluckman, P.: Scientific Advice in a Troubled World (2017). http://www.pmcsa.org.nz/blog/scientific-advice-in-a-troubled-world/

19. Graves, H.: Rhetoric in(to) science: Style as invention in inquiry. Cresskill, NJ: Hampton Press, Inc. (2005)

20. Graves, H.: The rhetoric of (interdisciplinary) science: visuals and the construction of facts in nanotechnology. Poroi 10(2). (2014). https://doi.org/10.13008/2151-2957.1207

21. Guston, D.H.: Boundary organizations in environmental policy and science: an introduction. Sci. Technol. Human Values **26**(4), 399–408 (2001)

22. Jasanoff, S.: Science and Public Reason. Routledge, London, UK (2012)

23. Latour, B., Woolgar, S.: Laboratory Life: The Construction of Scientific Facts. Princeton University Press, Princeton, NJ (1986)

24. Leith, P., Vanclay, F.: Translating science to benefit diverse publics: engagement pathways for linking climate risk, uncertainty, and agricultural identities. Sci. Technol. Human Values **40**(6), 939–964 (2015)

25. Leith, P., Haward, M., Rees, C., Ogier, E.: Success and evolution of a boundary organization. Sci. Technol. Human Values **41**(3), 375–401 (2016)

26. Linell, P.: Discourse across boundaries: on recontextualizations and the blending of voices in professional discourse. Texts **18**(2), 143–157 (1998)

27. Luzón, M.J.: Public communication of science in blogs: recontextualizing scientific discourse for a diversified audience. Writ. Commun. **30**(4), 428–457 (2013)

28. Mehlenbacher, A.R.: Science Communication Online: Engaging Experts and Publics on the Internet. The Ohio State University Press, Columbus, OH (2019)

29. Meyer, M.: The rise of the knowledge broker. Sci. Commun. **32**(1), 118–127 (2010)

30. Prelli, L.J.: A Rhetoric of Science: Inventing Scientific Discourse. University of South Carolina Press, Columbia, SC (1989)

31. Rachul, C.: Digesting data: tracing chromosomal imprint of scientific evidence through the development and use of Canadian dietary guidelines. J. Bus. Tech. Commun. **33**(1), 26–59 (2019)

32. Ravotas, D., Berkenkotter, C.: Voices in the text: the uses of reported speech in a psychotherapist's notes and initial assessments. Text **18**(2), 211–240 (1998). https://doi.org/10.1515/text.1.1998.18.2.211

33. Rowley-Jolivet, E., Carter-Thomas, S.: Scholarly soundbites: Audiovisual innovations in digital science and their implications for genre evolution. In: Pêrez-Llantada, C., Luzón, M.J. (eds.) Science Communication on the Internet: Old Genres Meet New Genres, pp. 81–106. John Benjamins Publishing Company, Amsterdam (2019)

34. Saldaña, J.: The coding manual for qualitative researchers (3rd ed.). London, UK: Sage Publications, (2016)

35. Sarangi, S.: On demarcating the space between 'lay expertise' and 'expert laity.' Text **21**(1/2), 3–11 (2001). https://doi.org/10.1515/text.1.21.1-2.3

36. Smart, G.: Discourse coalitions, science blogs and the public debate on global climate change. In: Bawarshi, A., Reiff, M.J. (eds.) Genre and the Performance of Publics, pp. 157–177. Utah State University Press, Logan (2016)

37. Smart, G., Falconer, M.: The representation of science and technology in *Laudato Si*: The role of a Vatican genre set in developing knowledge and argumentation. In: Pêrez-Llantada, C., Luzón, M.J. (eds.) Science Communication on the Internet: Old Genres Meet New Genres, pp. 195–217. John Benjamins Publishing Company, Amsterdam (2019)

38. Smart, G., Falconer, M.: The uptake and recontextualization of climate-change science within 'denialist' cultural communities. In: Auken, S., Sunesen, C. (eds.) Genres of the Climate Debate, pp. 85–107. De Gruyter, Warsaw/Berlin (2021)
39. Smart, G., Currie, S., Falconer, M.: Research on knowledge-making in professional discourses: The use of theoretical resources. In: Bhatia, V., Bremner, S. (eds.) The Routledge Handbook of Language and Professional Communication, pp. 85–98. Routledge, London and New York (2014)
40. Star, S.L.: This is not a boundary object: reflections on the origin of a concept. Sci. Technol. Human Values **35**(5), 301–317 (2010)
41. Star, S.L., Gruesemer, J.R.: Institutional ecology, "translations" and boundary objects: amateurs and professionals in Berkeley's Museum of Vertebrate Zoology, 1907–1939. Soc. Stud. Sci. **19**, 387–420 (1989)
42. Walker, K.C.: Mapping the contours of translation: visualized un/certainties in the ozone hole controversy. Tech. Commun. Q. **25**(2), 104–120 (2016)

Matthew A. Falconer is an Adjunct Research Professor in Carleton University's School of Linguistics and Language Studies. His research explores the uses of science in non-scientific domains. His doctoral research involved an ethnography of the Council of Canadian Academies' collaborative discursive boundary work of transforming science for policy-makers. He has published on the uses and misuses of science in the ongoing public climate change debate, and knowledge-making practices in science.

A Conceptual Framework for Knowledge Exchange in a Wildland Fire Research and Practice Context

Colin B. McFayden, Lynn M. Johnston, Douglas G. Woolford, Colleen George, Den Boychuk, Daniel Johnston, B. Mike Wotton, and Joshua M. Johnston

Abstract Wildland fire is an important natural disturbance in many vegetated areas of the world. However, fire management actions are critical not only to prevent and suppress unwanted fires, but also mitigate and recover from the negative impacts of fire on people and communities. Advancements in wildland fire science can help inform these necessary actions in wildland fire management. How science is created and integrated into these fire management decision-making processes, whether

C. B. McFayden (✉)
Ontario Ministry of Natural Resources and Forestry, Aviation, Forest Fire and Emergency Services, Dryden Fire Management Centre, Dryden, ON, Canada
e-mail: colin.mcfayden@nrcan-rncan.gc.ca

L. M. Johnston · B. M. Wotton · J. M. Johnston
Natural Resources Canada, Canadian Forest Service, Great Lakes Forestry Centre, Sault Ste. Marie, ON, Canada
e-mail: Lynn.Johnston@NRCan-RNCan.gc.ca

B. M. Wotton
e-mail: mike.wotton@utoronto.ca

J. M. Johnston
e-mail: joshua.johnston@NRCan-RNCan.gc.ca

D. G. Woolford
Department of Statistical and Actuarial Sciences, University of Western Ontario, London, ON, Canada
e-mail: dwoolfor@uwo.ca

C. George
Ontario Ministry Natural Resources and Forestry, Science and Research Branch, Centre for Northern Forest Ecosystem Research, Thunder Bay, ON, Canada
e-mail: Colleen.George@ontario.ca

D. Boychuk · D. Johnston
Ontario Ministry of Natural Resources and Forestry, Aviation Forest Fire and Emergency Services, Sault Ste. Marie, ON, Canada
e-mail: den.boychuk@ontario.ca

D. Johnston
e-mail: dan.johnston@ontario.ca

© The Author(s), under exclusive license to Springer Nature Switzerland AG 2023
D. G. Woolford et al. (eds.), *Applied Data Science*, Studies in Big Data 125,
https://doi.org/10.1007/978-3-031-29937-7_12

through collaborations with external researchers and/or with scientists within a wildland fire management agency itself, requires a conscious understanding of how the science is useful and goes beyond the simple existence of knowledge. This chapter outlines the goal of integrating fire science and management using a conceptual knowledge exchange (KE) framework, informed from existing work on KE. We provide a review of the KE literature relevant to wildland fire management and develop a KE framework for the fire management context. In this context, we address the potential barriers and facilitators throughout this process followed by a discussion of an active learning approach aimed at developing effective data translation skills amongst students in a data analytics consulting course.

Keywords Wildland fire · Knowledge exchange · Knowledge transfer · Technical transfer · Fire management · Fire science

1 Introduction

The location, time, size, and intensity of wildland fires are highly variable, and the impacts of these fires are often complex. Wildland fire can be beneficial, playing a role in the natural functioning of many fire adapted and fire dependent ecosystems while also reducing hazardous fuels [6]. However, wildland fire can have catastrophic outcomes for human communities, including loss of life, evacuations, and socio-economic disruptions [17]. For example, a single fire in Fort McMurray, Alberta, Canada in 2016 resulted in billions of dollars in insured losses along with considerable but unquantified impacts on families and first responders [26].

Wildland fire management is critically important for reducing negative impacts of wildland fire [7, 20]. It commonly focuses on suppression but often includes prevention, mitigation, and recovery [17, 31, 45]. The objectives typically emphasize protection of people, property, infrastructure, forest resources and socio-economic activity [45].

Fire management is very expensive. Over $1B can be spent annually on fire management in Canada [15, 38]. Fire management is also challenging and complex, involving decision-making across a wide range of spatial and temporal scales [2], high uncertainty, and multiple conflicting objectives. Operational fire management must deal with relatively infrequent but critical situations of extreme and quickly changing fire behavior and workloads, dangerous working conditions, and severe resource shortages.

Advancements in wildland fire science can help inform wildland fire management. Wildland fire science is both a body of *knowledge*[1] and a systematic process to build

B. M. Wotton
Graduate Department of Forestry, John. H. Daniels Faculty of Architecture, Landscape and Design, University of Toronto, Toronto, ON, Canada

[1] **Knowledge** can be classified into explicit (for example codified) and tacit knowledge (for example has a personal quality) [28]. Knowledge and knowledge creation occur over a range of domains from

and organize knowledge about wildland fire including topics that pertain to the needs of fire management. Wildland fire science occurs across a broad range of domains, approaches, and scales and their interactions. Such research has occurred over many decades (e.g., [6, 55]). This work is crucial for effective, efficient, and robust fire management as noted by Sankey's [37] recent blueprint for wildland fire science that outlined the need for both continued and new research to further the understanding of wildland fire in Canada.

An increasingly important area for fire science knowledge is wildland fire and climate change interactions. Existing research has shown how fire management in Canada may change under a range of possible future climates. For example, forest fuels are expected to be drier and, therefore, more receptive to ignition and vigorous fire spread. These factors are expected to result in having more and larger fires that exceed limits of direct suppression [10, 53, 54]. Studies on the effect of these changes on fire management in Ontario, Canada have shown that increases in fire occurrence and behavior compound non-linearly to an even greater proportion of escaped fires [51], requiring an even greater number of suppression resources [52].

Notwithstanding the many successful applications of science in fire management, developing and integrating science is not straightforward, nor without difficulties. The existence of knowledge itself is not sufficient to create a change in policies and practices [19, 33]. How science is shaped and used is complex and can, for example, be discovery or mission-driven with important interactions with policy contexts [29]. Science knowledge cannot be easily transferred and taken up by fire management agencies without addressing multiple factors that influence integration, including the relevance, credibility, and accessibility of the science and the operational, administrative, and cultural state of agencies [16, 19]. How science is created and integrated into these fire management decision-making processes, whether through collaborations with external researchers and sometimes even with scientists within a wildland fire management agency itself, requires a conscious understanding of how the science is useful. Focusing solely on identifying science gaps or improving communications between *researchers*[2] and *practitioners*[3] in disciplinary silos can be somewhat effective, but is limited if not done in an interdisciplinary, informed, collaborative, and

---

fundamental research to local communities [34]. It is important to recognize that the knowledge systems described here are derived from Western perspectives. The authors acknowledge the value of Indigenous and traditional ways of knowing and of knowledge exchange that are not represented in this paper. Indigenous ways of knowing celebrate the intimate connections between humans and the biophysical world. Fire has been used as an important tool for Indigenous Peoples for a variety of reasons, including in hunting and gathering activities, to regenerate land and safeguard resources, for cooking, heating, and ceremony, and for communication [24]. Indigenous Peoples hold important place-based knowledge about fire and fire management and have played a key role in wildland fire management through time.

[2] A **researcher** is a person who studies a subject and carries out academic or scientific research especially in order to discover new information or reach a new understanding (for example, a fire research scientist) (adapted from Cambridge Dictionary [4]).

[3] A **practitioner** is a person actively engaged in a discipline, or practices a profession for example, fire management staff, personnel, or managers [23].

iterative way [43]. We also recognize that this important design task can be aided using a *knowledge exchange* (KE) framework.

Effective KE in fire management helps ensure that real-world problems are understood by researchers, the research is relevant, and the results are integrated into fire management practices. This chapter outlines an example of a general conceptual KE framework informed from literature, and in this case is applicable to fire management. This KE framework supports the creation of application-oriented science outcomes and their successful adoption into operational fire management decision-making. We provide a review of the KE literature relevant to wildland fire management. Through developing a KE framework for the fire management context, we: (1) support the implementation of science *innovations*[4] into fire management agencies and (2) identify potential barriers and facilitators to KE in this context. We conclude with a discussion of an active learning approach that we have employed in a data analytics consulting course to help students with strong technical backgrounds in statistics, and data science and analytics develop their data translation skills.

2 Knowledge Exchange (KE)

There is no universal framework for KE. Concepts and terminology vary depending on both the domain and the focus (e.g., see [13, 14, 19, 25, 34, 35, 48]). In the literature, and in everyday use, there are many terms that are used interchangeably or with different meanings and are elaborated further in the cited references. These terms include knowledge translation [39], knowledge mobilization [19], knowledge to action [14] and knowledge transfer [12].

We define KE as: (1) the collective overarching process where knowledge is collaboratively created, shared and transformed; and (2) the context for learning about new knowledge [18, 33, 34]. KE implies feedback within a network of researchers, intermediaries, and practitioners [9].

KE has been described as a system where reciprocal learning to discover, create, or address something with mutual understanding and benefit can occur [33, 35]. Through this understanding of KE, outcomes tend to be more realistic, acceptable, and likely to produce more lasting change [35].

It is crucially important to emphasize that we consider KE as an iterative process with bi-directional flows [9, 33, 34] where researchers and practitioners are both knowledge producers and users. This contrasts with typical historical practice where researchers push and practitioners pull knowledge between the two groups. Our experience suggests those one-way knowledge streams from producers to users are insufficient in the wildland fire community, rather, shared understanding and concerted

---

[4] **Innovation** is the adoption of the products and related organizational, administrative or policies related to fire management agencies (adapted from Damanpour and Gopalakrishnan [8]. In this way innovation is viewed as an outcome of knowledge exchange. Adoption is synonymous with implementation and integration.

efforts to create and diffuse knowledge are needed [3, 43]. Early and continuous knowledge flow between practitioners and researchers has been shown to be an essential approach for KE in wildland fire. For example, Woolford et al. [50] noted how instrumental this type of knowledge flow was in the development and implementation of a province-wide, fine scale, spatially explicit human-caused wildland fire occurrence mode for Ontario, Canada. KE is a complex non-linear process, with interactions between sub-systems [9, 14], KE can be conceptualized as a network or web as illustrated in Fig. 1, which is an example with several networks of researchers or research groups and practitioners or practitioner groups in different domains, all working to identify and address a specific problem.

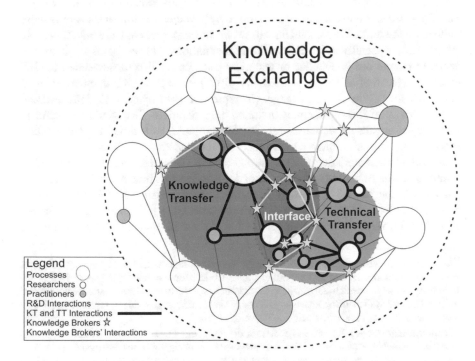

Fig. 1 Conceptual illustration of knowledge exchange (KE) and its knowledge transfer (KT) and technical transfer (TT) sub-processes for a specific, hypothetical case of science research and development (R&D) and integration. KE is an overarching process among researchers and practitioners. The sizes of the researchers' and practitioners' circles represent their respective levels of expertise for this specific case. The black lines represent connections among people during the R&D and integration (KT, TT) work. The thick circles identify the people involved in the KT and TT sub-processes. The thick black lines represent connections between people for the KT and TT work. The yellow stars represent knowledge brokers, who facilitate connections among various people and groups. The yellow lines represent connections between knowledge brokers. The interface of KT and TT represents the interactions between researchers and practitioners that seek to increase their respective and mutual understanding. Defined boundaries are shown for the interface between KT and TT, but the actual boundaries are a fuzzy continuum

Once an applied outcome becomes clearer, the efforts transition to *knowledge*[5] *and technical transfer*[6] processes. *Knowledge brokers*[7] support this exchange at all stages, facilitating collaboration and bridging knowledge between researchers, practitioners. The interface between knowledge and technical transfer conceptually has the highest concentration of knowledge brokers because this is where 'the water hits the fire' so to speak and innovation and implementation take place. *Innovation*[8] can have several meanings and take various forms (e.g., a system, idea, or tool). We consider the outcome of KE as evidence-informed application of scientific innovation aimed at achieving a specific outcome for fire management policies or practices.

Gopalakrishnan and Santoro [13] proposed that knowledge and technical transfer are different in scope and facilitated by different organizational factors. Knowledge transfer is broader and concerned with the 'why', whereas technical transfer is more focused on the tools. We contend that these two sub-processes of KE work together in many cases, especially in novel or unfamiliar situations. The attributes and activities needed to carry out knowledge or technical transfer are like those needed for KE in general. Reed et al. [33] identified five directives to guide KE in environmental management: (1) design, (2) engage, (3) represent, (4) impact, and (5) reflect and sustain. Many of these principles included considerations that would be useful to knowledge transfer or technical transfer, for example, well-timed implementation and creating networks suitable to the scope of the transfer.

A critical characteristic of KE is mutual benefit. This place where mutual under-standing, communications, sharing, and knowledge creation occurs is called the knowledge interface [34]. Loosely described, the *knowledge and technical inter-face*[9] is the place for collaboration—a critical aspect of KE. Roux et al. [34] further described the values of shared understanding, where participants move beyond the

---

[5] **Knowledge transfer** is a sub-process of KE for disseminating broader learning aimed at changes in strategic thinking, culture and providing inputs to decision-making [13]. This embodies the underlying principles which may include considering aspects such as organizational design and culture. This is a systematic approach to collect and share knowledge so ideas, research results and skills enable innovative new products to be developed [14].

[6] **Technical transfer** is a sub-process of KE for disseminating knowledge with a more narrow-in-focus than knowledge transfer and aimed at processes, products, tools, data or models [13]. This may include considering aspects such as policy, procedures for acquisition, application and archive of information [56].

[7] **Knowledge brokers** (data translators; opinion leaders, boundary organizations) are the inter-mediaries between the knowledge producers and those who use it. They are the human force behind finding, assessing and interpreting evidence, facilitating interaction and identifying emerging research questions [28, 29, 49]. Knowledge brokers may be specialized to certain domains such as data translators who bridge the expertise gaps between technical teams in data science [22]. Knowledge brokers may also be opinion leaders who are trusted information sources [3]. There are also boundary organizations which are coordinated groups that are intermediaries that develop long-term relationships and collaboration to increase the impact of science in fire management [16].

[8] See footnote 4.

[9] The **knowledge and technical interface** is where concerted bi-directional flow of collaborative learning, shared understanding of key concepts and co-evolution towards common purpose, intent and action takes place [34]. We contend this is where tacit and explicit knowledge exchange can be the most impactful and therefore important for the positioning of knowledge brokers.

typical role of knowledge producer and user and negotiate what is achievable and relevant. This interface (or collaboration) facilitates increased quality and acceptance of science because of trust and aligned incentive systems.

3 Knowledge, Researchers and Practitioners

Research is investigation in a planned and systematic fashion for the purpose of increasing the sum of knowledge [29], typically done by a researcher or research team. Within a fire management agency context, a community of practice can refer to those who manage an aspect of fire, such as a cadre of Fire Behaviour Analysts or firefighters. We can refer to these people as practitioners, as suggested by McGee et al. [23]. The creation and holding of knowledge occurs across five generalized domains, which have different degrees of the explicitness of knowledge [34]. Figure 2 summarizes this continuum with examples pertaining to fire behaviour. Although presented as distinct and separate, we recognize the boundaries between knowledge domains are fuzzy, and there are individuals whose expertise span multiple domains (such as a researcher who is also a practitioner). We also recognize that knowledge comes in many forms from Indigenous Knowledge to experiential and operational knowledge [43].

The knowledge being exchanged can range from more formal knowledge, known as "explicit knowledge", to knowledge that is more subjective and based on ideas,

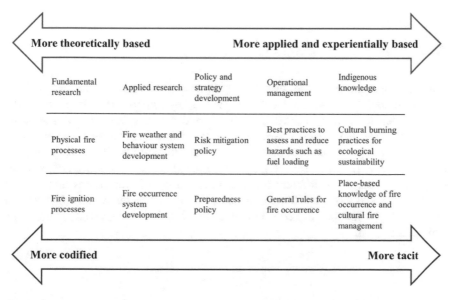

Fig. 2 Examples of knowledge domains adapted from [34] in a wildland fire behaviour context. The degree of codified knowledge is not the same as the degree of rules for policy implementation

perceptions, or experience, known as "tacit knowledge" [1]. Explicit knowledge is more easily expressed and codified, whereas tacit knowledge is more subtle and often difficult to convey. The assumption is that shared contexts and understanding in respective knowledge domains results in arguably better outcomes for both parties [35]. This facilitates acceptance, sustained use, and growth of the knowledge [34].

4 The Role of Knowledge Exchange from Problem Identification to Implementation

Having established context and elements of KE as an overarching system, the focus turns to the sub-processes that bridge knowledge creation through to implementation. Graham et al. [14] visualized this as a cycle. We build on these ideas (Fig. 3) and place the cycle in a fire management context. It is important first to note that knowledge and technical transfer occur at varying times and with varying complexity in the journey from problem identification through knowledge creation to implementation.

How does KE happen? These processes are aided by knowledge brokers to encourage and facilitate positive interactions at the knowledge interface (as visualized conceptually in Fig. 1). It starts with having the right people in that knowledge and technical interface space who recognize a problem or research need. The remainder of this section describes the system and processes underlying KE as illustrated in Fig. 3. We describe the application of KE to wildland fire management, although the system and process are more widely applicable.

4.1 Problem Identification

There is not a single person, group, or path to achieve science-informed policies and practices for fire management; however, a key requirement for effective and efficient development of relevant, practical, and useful science is to have some individuals who have deep expertise in both science and fire management. This is essential for (1) understanding problems correctly and identifying opportunities where currently feasible research may help, and (2) ensuring effective communication among people from different domains. Problem identification spans domains and can be facilitated through different avenues; examples include formal collaboration agreements, memoranda of understandings, and informal professional relationships and participation.

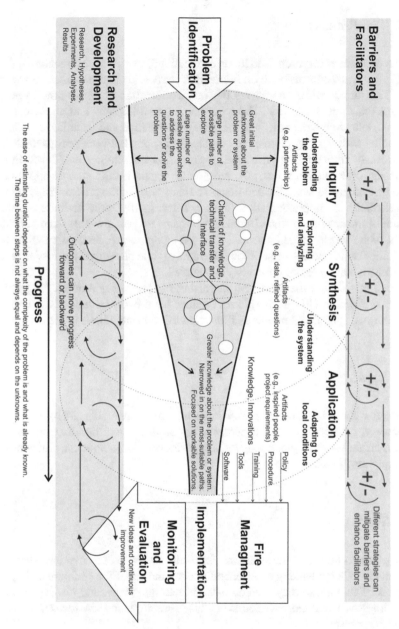

Fig. 3 Illustration of the systems and processes of knowledge exchange towards addressing problems and advancing innovation for fire management. The inquiry and synthesis elements may be widely applicable, while the application, implementation, monitoring and evaluation aspects are specific to our fire management experience

4.2 The Process

Once a problem or research need has been identified, work can commence to address it. This process is illustrated by a funnel that starts wide and becomes narrow over time. The funnel width represents the relative uncertainty about how to address the problem in current work. There are many possible paths and approaches (for example, knowledge domains and methods). Knowledge and technical transfer and interfacing with varying degrees of complexity happen between groups throughout this process. The funnel narrows with progress as uncertainty is reduced through knowledge creation and access; the most-suitable path becomes more apparent and the focus changes to workable solutions. The narrowing of the process from 'problem identification' to 'implementation' does not necessarily indicate there is a lower amount of uncertainty or the elimination of paths and approaches from consideration in future work. Indeed, focused work on complex problems often allow more recognition and quantification of the uncertainty while also revealing new aspects of the problem that were not previously considered.

Along this funnel there are the interacting phases of (1) inquiry, (2) synthesis and (3) application [14]. These phases interact, have fuzzy boundaries, and overlap depending on specific situations as illustrated using hashed lines in Fig. 3.

4.3 Inquiry

The inquiry phase is characterized by the many options available and by exploration, uncertainty, creation of desired or necessary skillsets, and building partnerships. Process artifacts of this phase may include partnerships, agreements, brainstorming, and exploratory data.

4.4 Synthesis

As progress continues to the synthesis phase, the focus shifts to making sense of the relevant knowledge leading to a general understanding of the problem and system. Artifacts of this phase may include data, refined questions, and discrete work. As clarity improves, and more workable outcomes are produced, the focus moves to the application phase.

4.5 Application

In this phase, some innovation or knowledge is suitable for application in fire management. There may be efforts to adapt outcomes to local conditions for a variety of potential audiences or purposes across the fire management field. Artifacts in this stage include inspired people, codified knowledge, requirements for other implementation processes, a new or amended policy, an improved procedure, a new tool, or prototype software.

The fire management decision-making space is complex [2, 41, 42, 44, 56]. In addition, fire management is very user focused. The specifics of how a new idea or product is developed should be aligned to the end user needs and the decision-making environment [18]. This is not always straightforward because there are many complex challenges for fire management that occur at different scales and scopes, from real-time decisions on a single fire to longer-term, national-level policy setting [40, 45].

Successful application requires the effective interaction between researchers and practitioners for translation, support, and delivery of the necessary knowledge [23, 25, 36]. Within the wildland fire community, early and ongoing close engagement between researchers and practitioners is critical to successful decision support system development and implementation because of the need for shared understanding [21, 27, 50].

4.6 Implementation

Implementation almost always requires a tailored solution. There are specific ways that the outcomes of the KE process can be implemented, such as a policy review cycle, procedure task team, or project plan. However, given the context of the public sector where most fire management agencies in Canada are positioned [5], innovation is often challenging because it can be seen as unknown in an organizational structure that discourages risk [30]. Although multiple agencies may share common fire management problems, factors such as different governance structures and practices can slow or impede innovation. Regulatory or legal constraints may also slow or impeded innovation in codified practices. Adoption by practitioners through passive dissemination can sometimes be ineffective [49]. We view knowledge and technical transfer as a sub-process as distinct from project planning or software development methods. The latter are commonly used as mechanisms to manage the creation of initially relatively well-defined projects or products such as training courses or software (e.g., [46]). Project planning approaches are appropriate for the application phase of Fig. 3 when the problem and solution are well understood. It is very important to understand which implementation method is needed based on the fire management agency institutional requirement. After implementation, monitoring and continued evaluation should occur and may result in new ideas for future work. This is a practice of continuous improvement.

4.7 Processes of Progression and Retrogression

There are two parallel considerations that are pervasive throughout KE and influence progress at all phases. These are (1) the research and development process and (2) barriers and facilitators. These are illustrated above and below the funnel in Fig. 3.

4.8 Research and Development Cycle

The research and development process includes exploration, discovery, trial and error, hypothesis testing, confirmation, prototyping, and field testing. This necessarily involves advancing and retreating as tentative results emerge. This cycling tends to occur earlier but can happen at any point. This moves us forward and back in the funnel in larger or smaller steps.

4.9 Barriers and Facilitators

Identifying and understanding the significance of barriers to and facilitators of progress are critical within fire management agencies. This is true for both KE, where the focus is between researchers and practitioners, and knowledge and technical transfer, where the focus is on the needs of the adopter. Other areas such as health sciences and conservation are further along with KE research and the identification of barriers and facilitators [25, 48] and recently conversations associated with KE have arisen in the wildland fire management literature [16, 43].

There are many potential categories of barriers and facilitators, and we need a tractable way to understand them. We compared barriers and facilitators identified from wildland fire science centric KE papers [9, 23, 36] and three recent perspectives on the adoption of wildland fire decision support [21, 27, 32]. We organized the comparison commencing with the barriers and facilitators summary by Mitton et al. [25] wherein these were classified for policy decision-making for health studies. Figure 4 is a summary of the barriers and facilitators pulled and organized from the six wildland fire papers and organized by the recurring themes also considering those from Mitton et al. [25]. The nine themes of barriers and facilitators we use are capacity, clear objectives and alignment, collaboration and networking, communication, ownership and authority, readiness for innovation, research motivation, timing, and trust. Barriers and facilitators identified by the majority of authors (at least 3 of 6 papers noted above) include: limited time to make decisions; collaborative research

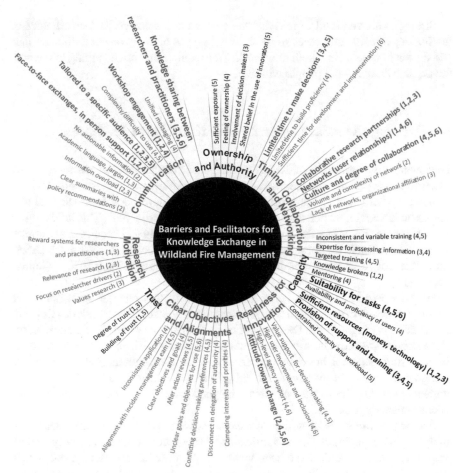

Fig. 4 Barriers and facilitators for knowledge exchange in wildland fire management. Nine central themes for the barriers and facilitators were identified in selected literature. Each barrier and facilitator identified in the literature is listed under a specific theme. Barriers and facilitators that were in at least half (≥3) of the papers are shown in a bold and larger font. The barriers and facilitators identified in the papers are numbered as follows: (1) McGee et al. [23]; (2) Davis et al. [9]; (3) Ryan and Cerveny [36]; (4) Noble and Paveglio [27]; (5) Rapp et al. [32]; (6) Martell [21]. Papers from the literature were selected from two general scopes: papers 1–3 describe knowledge exchange (researcher and practitioner); and papers 4–6 discuss knowledge and technical transfer (innovation and adopters)

partnerships; networks (user relationships); culture and degree of collaboration; suitability for task; sufficient resources (money, technology); provision of support and training; attitude towards change; face-to-face exchanges, in person support; tailored to a specific audience; workshop engagement; and knowledge sharing between researchers and practitioners.

Strategies are required to mitigate the barriers and enhance the performance of facilitators. Specific strategies must align with types of decisions practitioners face and the environments in which they work. This requires an understanding of the both the organization and the people within it.

5 Training the Next Generation in Knowledge Exchange

The overarching principle of KE as a mutual exchange between researchers and practitioners is perhaps best learned through experience. In the classroom, this can be achieved using active learning techniques, which have been found to lead to a deeper understanding when compared to traditional lecturing [47]. This holds true in the data science domain—the importance of active learning techniques was endorsed by the statistical science community in the American Statistical Association's (ASA) Guidelines for Assessment and Instruction in Statistics Education (GAISE) College Report [11].

We have explored the KE principles outlined in this chapter using an active learning approach in the context of a post-secondary course, "Data Analytics Consulting", which is taught in the Department of Statistical and Actuarial Sciences at the University of Western Ontario (https://www.uwo.ca/stats/). This course is offered to 4th-year students in the honours Statistical or Data Science undergraduate programs and graduate students (Masters and PhD) in that department, as well as graduate students pursuing the Master of Data Analytics program, a one-year professional science master's program.

Although those students will have received advanced training in data science and analytics theory, techniques, and applications through data modelling, nearly all their preceding training would have been technical in nature. Consequently, rather than teaching new data modelling theory or techniques, the course's learning objectives focus on fostering the development of key skills to be a successful data science and analytics professional, including developing skills necessary for effective data translation. A variety of topics are covered, such as: the iterative flow through the data analytics consulting process; meetings and project management; intellectual property, compensation, and negotiation; robust and ethical data analyses. These are all grounded in the development of effective communication skills, needed for KE, which is threaded throughout the course content and its assessments.

Active learning is incorporated through a community engaged learning approach, where an external "client" (not the instructor or a teaching assistant) interacts with the class throughout the term. Students are grouped into teams and, through a series of interactions with the client that take place over the course of the term, they practice and develop their KE skills. Those interactions mimic typical engagement settings, including synchronous and asynchronous learning opportunities. Examples of synchronous engagement include an initial meeting, phone/video meetings, interim presentation(s) and discussions, as well as an in-person meeting (when feasible). Asynchronous activities include communicating via email, providing, and receiving feedback on interim progress report(s). In all such interactions, the instructor acts as a knowledge broker, facilitating interactions between the students and the client while also having separate interactions with the client where they discuss how to help guide the students through this KE process. A final report to the client, written in appropriate format and language for that target audience is used as a final summative assessment. All of this occurs using a directed learning approach where both the client and the instructor act as knowledge brokers to guide the students through this process. Informal peer assessments are also included for some engagement pieces so that the students can observe and learn from their fellow classmates while also providing constructive criticism. A sample schedule of activities for a 4-month term appears in Table 1.

In essence, this active learning using a community engaged learning approach for the Data Analytics Consulting course both teaches and applies KE principles. The classroom is a place for knowledge interface where fire management practitioners (one of the clients for the past few years), data scientist and the students interact for mutual benefit. The students learn from the practitioners about their domain and gain a deeper understanding of the meaning of the data. The practitioners learn new

Table 1 Sample schedule of knowledge exchange activities in the data analytics consulting class

Week (s)	Topic	Activities
1	Initial meeting (virtual) and problem description	Presentation by the client
2	Data, data governance and project overview	Data sharing agreements signed Data released Teams identified
3	Exploratory data analysis presentations and discussion	5 min presentations by each team Planning for next touchpoint with client
4	Client meeting (virtual)	Remote synchronous meeting to discuss results of exploratory data analysis and ask any questions
5	Preliminary modelling	Planning for client's visit the following week
6	Client meetings (in person)	Teams present and discuss summaries of work to date with client Debriefing after meetings, led by instructor with peer discussions and feedback
7–13	Ongoing project work and client engagement	Team presentations Class discussions with peer feedback Client meetings (virtual; approx. bi-weekly) Final presentations
14–16	Community engaged learning project ends	Final written reports submitted Client provides feedback, which is incorporated into each team's grade

ways data can be informative in their business. These interactions lead to better understanding, new initiatives, and importantly, inspired people, improving both the fire practitioner's knowledge of data science and its application and the student's knowledge of fire management.

This is one approach that can help meet the bi-directional flow of KE while also recognizing in the academic environment that there is a need to develop those communication, business, and soft skills to become a knowledge broker. That is, to become an effective data translator, working in the wildland fire science and management domain. The outcomes of these efforts were presented at Wildland Fire Canada 2019, which is part of a biennial series of conferences (https://wildlandfire canada.com/) that bring together a wide variety of people working in wildland fire, both fire management practitioners and wildland fire science researchers.

Finally, it is important to note that similar approaches can be applied to effectively train students outside of a classroom setting when they are conducting thesis-based research guided by a supervisor. In this context, regular engagement between the student, other researchers, and practitioners is crucial to foster the development of effective KE skills. Supervisors can act as knowledge brokers, encouraging the student to not only attend and participate in such interactions, but to also have them witness the interactions of other trainees in the research lab to learn from their peers. These interactions also help the trainees expand their professional network.

6 Closing

Fire management is challenging and will become even more so in future. Globally, a large and growing amount of wildland fire science work is being done to aid fire management. The integration of ongoing advances remains difficult and occurs slowly, which can leave fire management understanding and practices short of the best available science and necessary innovation. Research efforts can fall short of effective implementation and typically end with traditional, impersonal approaches such as publications and reports [19] and other studies have identified barriers in the public service's use of research [29].

Ultimately, people and relationships are a crucial vehicle for overcoming barriers to successfully integrating science into practice. KE is not about processes and check-lists. People are at the heart of KE, and effective KE depends on networks of diverse people and teams in which individuals can play one or more roles. These individuals need not only technical skills, but also creativity and soft social skills [30]. It takes significant effort on the part of agencies to engender and successfully integrate new science into operational fire management practices and decision-making. Similarly, it takes extra effort for researchers and practitioners to maintain strong working relationships even though they are mutually beneficial. Our repeated experience as fire management practitioners, researchers and knowledge brokers suggests that these efforts are exceptionally beneficial for fire management and rewarding for all concerned.

While fire management agencies have practiced some elements of KE for years, adoption of holistic KE thinking is relatively recent and continues to improve. There is no single, authoritative KE system and process. In this chapter, we attempted to organize KE components into a framework to support the implementation of science innovations in the wildland fire management context; although, the framework offers value to other contexts. Our ongoing KE work involves developing more detailed, practical guidance for KE and application of KE for fire management innovations.

Acknowledgements This work was completed in part to support the WildFireSat mission fire management engagement planning. We also acknowledge the support of the Natural Sciences and Engineering Research Council of Canada (NSERC) and the Ministry Natural Resources and Forestry. We thank Meghan Sloane for technical support and assistance with the literature review. We also thank the Editors and reviewer for their helpful comments.

References

1. Bolisani, E., Scarso, E.: Information technology management: a knowledge-based perspective. Technovation **19**(4), 209–217 (1999)
2. Boychuk, D., McFayden, C.B., Evens, J., Shields, J., Stacey, A., Woolford, D.G., McLarty, D.: Assembling and customizing multiple fire weather forecasts for burn probability and other fire management applications in Ontario, Canada. Fire **3**(2), 16 (2020). https://doi.org/10.3390/fire3020016
3. Butler, B.W., Brown, S., Wright, V., Black, A.: Bridging the divide between fire safety research and fighting fire safely: how do we convey research innovation to contribute more effectively to wildland firefighter safety? Int. J. Wildland Fire **26**(2), 107–112 (2017). https://doi.org/10.1071/WF16147
4. Cambridge Dictionary (2022) Researcher. Cambridge Dictionary. https://dictionary.cambridge.org/dictionary/english/researcher
5. Canadian Interagency Forest Fire Centre (2022) CIFFC Member agencies. Canadian Interagency Forest Fire Centre (CIFFC). https://www.ciffc.ca/fire-information/member-agencies
6. Coogan, S.C., Daniels, L.D., Boychuk, D., Burton, P.J., Flannigan, M.D., Gauthier, S., ... Wotton, B.M. (2021). Fifty years of wildland fire science in Canada. Can. J. For. Res. **51**(2), 283–302. https://doi.org/10.1139/cjfr-2020-0314
7. Cumming, S.G.: Effective fire suppression in boreal forests. Can. J. For. Res. **35**(4), 772–786 (2005). https://doi.org/10.1139/x04-174
8. Damanpour, F., Gopalakrishnan, S.: Theories of organizational structure and innovation adoption: the role of environmental change. J. Eng. Tech. Manage. **15**(1), 1–24 (1998). https://doi.org/10.1016/S0923-4748(97)00029-5
9. Davis, E.J., Moseley, C., Olsen, C., Abrams, J., Creighton, J.: Diversity and dynamism of fire science user needs. J. For. **111**(2), 101–107 (2013). https://doi.org/10.5849/jof.12-037
10. Flannigan, M.D., Logan, K.A., Amiro, B.D., Skinner, W.R., Stocks, B.J.: Future area burned in Canada. Clim. Change **72**(1), 1–16 (2005)
11. GAISE College Report ASA Revision Committee: Guidelines for assessment and instruction in statistics education college report 2016 (2016). http://www.amstat.org/education/gaise
12. Gilbert, M., Cordey-Hayes, M.: Understanding the process of knowledge transfer to achieve successful technological innovation. Technovation **16**(6), 301–312 (1996). https://doi.org/10.1016/0166-4972(96)00012-0
13. Gopalakrishnan, S., Santoro, M.D.: Distinguishing between knowledge transfer and technology transfer activities: the role of key organizational factors. IEEE Trans. Eng. Manage. **51**(1), 57–69 (2004). https://doi.org/10.1109/TEM.2003.822461

14. Graham, I.D., Logan, J., Harrison, M.B., Straus, S.E., Tetroe, J., Caswell, W., Robinson, N.: Lost in knowledge translation: time for a map? J. Cont. Educ. Health Prof. **26**(1), 13–24 (2006). https://doi.org/10.1002/chp.47

15. Hope, E.S., McKenney, D.W., Pedlar, J.H., Stocks, B.J., Gauthier, S.: Wildfire suppression costs for Canada under a changing climate. PLoS ONE **11**(8), e0157425 (2016). https://doi.org/10.1371/journal.pone.0157425

16. Hunter, M.E., Colavito, M.M., Wright, V.: The use of science in wildland fire management: a review of barriers and facilitators. Curr. For. Rep. **6**, 1–14 (2020)

17. Johnston, L.M., Wang, X., Erni, S., Taylor, S.W., McFayden, C.B., Oliver, J.A., ... Flannigan, M.D.: Wildland fire risk research in Canada. Environ. Rev. **28**(2), 164–186 (2020). https://doi.org/10.1139/er-2019-0046

18. Lavis, J.N., Robertson, D., Woodside, J.M., McLeod, C.B., Abelson, J.: How can research organizations more effectively transfer research knowledge to decision makers? Milbank Q. **81**(2), 221–248 (2003). https://doi.org/10.1111/1468-0009.t01-1-00052

19. Levin, B.: Thinking about knowledge mobilization. In: An Invitational Symposium Sponsored by the Canadian Council on Learning and the Social Sciences and Humanities Research Council of Canada, pp 15–18 (2008)

20. Martell, D.L., Sun, H.: The impact of fire suppression, vegetation, and weather on the area burned by lightning-caused forest fires in Ontario. Can. J. For. Res. **38**(6), 1547–1563 (2008). https://doi.org/10.1139/X07-210

21. Martell, D.: The development and implementation of forest fire management decision support systems in Ontario, Canada: personal reflections on past practices and emerging challenges. Math. Comput. For. Nat. Res. Sci. **3**(1) (2011)

22. Maynard-Atem, L., Ludford, B.: The rise of the data translator. Impact **2020**(1), 12–14 (2020). https://doi.org/10.1080/2058802X.2020.1735794

23. McGee, T.K., Curtis, A., McFarlane, B.L., Shindler, B., Christianson, A., Olsen, C., McCaffrey, S.M.: Facilitating knowledge transfer between researchers and wildfire practitioners about trust: an international case study. For. Chron. **92**(2), 167–171 (2016). https://doi.org/10.5558/tfc2016-035

24. McKemey, M.B., Ens, E.J., Hunter, J.T., Ridges, M., Costello, O., Reid, N.C.: Co-producing a fire and seasons calendar to support renewed Indigenous cultural fire management. Austral Ecol. **46**(7), 1011–1029 (2021). https://doi.org/10.1111/aec.13034

25. Mitton, C., Adair, C.E., McKenzie, E., Patten, S.B., Perry, B.W.: Knowledge transfer and exchange: review and synthesis of the literature. Milbank Q. **85**(4), 729–768 (2007). https://doi.org/10.1111/j.1468-0009.2007.00506.x

26. MNP: A review of the 2016 Horse River Wildfire. Alberta agriculture and forestry preparedness and response. MNP LLP, Alberta Agriculture and Forestry (2017). https://www.alberta.ca/assets/documents/Wildfire-MNP-Report.pdf

27. Noble, P., Paveglio, T.B.: Exploring adoption of the wildland fire decision support system: end user perspectives. J. For. **118**(2), 154–171 (2020). https://doi.org/10.1093/jofore/fvz070

28. Nonaka, I.: A dynamic theory of organizational knowledge creation. Organ. Sci. **5**(1), 14–37 (1994). https://doi.org/10.1287/orsc.5.1.14

29. Nutley, S. M., Walter, I., Davies, H.T.: Using Evidence: How Research Can Inform Public Services. Policy Press (2007)

30. OECD: Fostering Innovation in the Public Sector. OECD Publishing, Paris (2017). https://doi.org/10.1787/9789264270879-en

31. OMNRF: Wildland fire management strategy. Ontario Ministry of Natural Resources and Forestry. Queen's Printer for Ontario, Toronto (2014)

32. Rapp, C., Rabung, E., Wilson, R., Toman, E.: Wildfire decision support tools: an exploratory study of use in the United States. Int. J. Wildland Fire **29**(7), 581–594 (2020). https://doi.org/10.1071/WF19131

33. Reed, M.S., Stringer, L.C., Fazey, I., Evely, A.C., Kruijsen, J.H.: Five principles for the practice of knowledge exchange in environmental management. J. Environ. Manage. **146**, 337–345 (2014). https://doi.org/10.1016/j.jenvman.2014.07.021

34. Roux, D.J., Rogers, K.H., Biggs, H.C., Ashton, P.J., Sergeant, A.: Bridging the science–management divide: moving from unidirectional knowledge transfer to knowledge interfacing and sharing. Ecol. Soc. **11**(1), 4 (2006). https://doi.org/10.5751/ES-01643-110104
35. Rushmer, R., Ward, V., Nguyen, T., Kuchenmüller, T.: Knowledge translation: key concepts, terms and activities. In: Verschuuren, M., van Oers, H. (eds.) Population Health Monitoring, pp. 127–150. Springer, Cham (2019). https://doi.org/10.1007/978-3-319-76562-4_7
36. Ryan, C.M., Cerveny, L.K.: Wildland fire science for management: federal fire manager information needs, sources, and uses. West. J. Appl. For. **26**(3), 126–132 (2011). https://doi.org/10.1093/wjaf/26.3.126
37. Sankey, S.: Blueprint for wildland fire science in Canada (2019–2029). Natural Resources Canada, Canadian Forest Service, Northern Forestry Centre (2018). http://cfs.nrcan.gc.ca/publications?id=39429
38. Stocks, B.J., Martell, D.L.: Forest fire management expenditures in Canada: 1970–2013. For. Chron. **92**(3), 298–306 (2016). https://doi.org/10.5558/tfc2016-056
39. Straus, S.E., Tetroe, J., Graham, I.: Defining knowledge translation. Can. Med. Assoc. J. **181**(3–4), 165–168 (2009). https://doi.org/10.1503/cmaj.081229
40. Taylor, S.W.: Wildfire management decision making—fast and slow: a systems framework for wildfire management research [Poster session]. In: 2017 Conference on Fire Prediction Across Scales. New York City, NY, United States (2017)
41. Taylor, S.W.: Atmospheric cascades shape wildfire activity and fire management decision spaces across scales—a conceptual framework for fire prediction. Front. Environ. Sci. **8**, 172 (2020). https://doi.org/10.3389/fenvs.2020.527278
42. Taylor, S.W., Woolford, D.G., Dean, C., Martell, D.L.: Wildfire prediction to inform fire management: statistical science challenges. Stat. Sci. **28**, 586–615 (2013). https://doi.org/10.1214/13-STS451
43. Tedim, F., McCaffrey, S., Leone, V., Vazquez-Varela, C., Depietri, Y., Buergelt, P., Lovreglio, R.: Supporting a shift in wildfire management from fighting fires to thriving with fires: the need for translational wildfire science. For. Policy Econ. **131**, 102565 (2021). https://doi.org/10.1016/j.forpol.2021.102565
44. Thompson, M.P., Calkin, D.E.: Uncertainty and risk in wildland fire management: a review. J. Environ. Manage. **92**(8), 1895–1909 (2011). https://doi.org/10.1016/j.jenvman.2011.03.015
45. Tymstra, C., Stocks, B.J., Cai, X., Flannigan, M.D.: Wildfire management in Canada: review, challenges and opportunities. Prog. Disaster Sci. **5**, 100045 (2020). https://doi.org/10.1016/j.pdisas.2019.100045
46. Varajão, J., Colomo-Palacios, R., Silva, H.: ISO 21500: 2012 and PMBoK 5 processes in information systems project management. Comput. Stan. Interfaces **50**, 216–222 (2017). https://doi.org/10.1016/j.csi.2016.09.007
47. Waldrop, M.M.: The science of teaching science. Nature **523**(7560), 272–274 (2015)
48. Walsh, J.C., Dicks, L.V., Raymond, C.M., Sutherland, W.J.: A typology of barriers and enablers of scientific evidence use in conservation practice. J. Environ. Manage. **250**, 109481 (2019). https://doi.org/10.1016/j.jenvman.2019.109481
49. Ward, V., House, A., Hamer, S.: Knowledge brokering: the missing link in the evidence to action chain? Evid. Policy J. Res. Debate Pract. **5**(3), 267–279 (2009). https://doi.org/10.1332/174426409X463811
50. Woolford, D.G., Martell, D.L., McFayden, C.B., Evens, J., Stacey, A., Wotton, B.M., Boychuk, D.: The development and implementation of a human-caused wildland fire occurrence prediction system for the province of Ontario, Canada. Can. J. For. Res. **51**(2), 303–325 (2021). https://doi.org/10.1139/cjfr-2020-0313
51. Wotton, M., Logan, K., McAlpine, R.: Climate change and the future fire environment in Ontario: fire occurrence and fire management impacts. In: Climate Change Research Report-Ontario Forest Research Institute (CCRR-01) (2005)
52. Wotton, B.M., Stocks, B.J.: Fire management in Canada: vulnerability and risk trends. In: Hirsch, K., Fuglem, P. (eds.) Canadian wildland fire strategy: background syntheses, analyses, and perspectives, pp. 49–55. Canadian Council of Forest Ministers, Natural Resources Canada (2006)

53. Wotton, B.M., Nock, C.A., Flannigan, M.D.: Forest fire occurrence and climate change in Canada. Int. J. Wildland Fire **19**(3), 253–271 (2010). https://doi.org/10.1071/WF09002
54. Wotton, B.M., Flannigan, M.D., Marshall, G.A.: Potential climate change impacts on fire intensity and key wildfire suppression thresholds in Canada. Environ. Res. Lett. **12**(9), 095003 (2017). https://doi.org/10.1088/1748-9326/aa7e6e
55. Wright, J.G.: Forest-fire Hazard tables for mixed red and white pine forests eastern Ontario and western Quebec regions. Department of the Interior Canada (1933)
56. Zimmerman, T.: Wildland fire management decision making. J. Agric. Sci. Technol. B **11**(2), 169–178 (2012)

Pedagogical and Future Implications for the Training of Data Translators

Boba Samuels, Donna Kotsopoulos, and Douglas G. Woolford

Abstract Sharing data stories is an exciting opportunity for integrating data into the workplace. The examples provided in the chapters of this book reveal three imperatives for training data translators: 1) interdisciplinarity, 2) a knowledge exchange framework (KEF), and 3) language calibration. We propose that a focus on developing students' skills in these areas must be coupled with data science training if we are to prepare effective data translators. Instructors must look to adjust not just their methods, but the philosophy behind their instruction to train students for the next generation of data translation jobs.

The sharing of data science stories within and beyond single disciplines presents an exciting opportunity that is increasingly recognized by data users within academia, workplaces of all types, and the public realm. In the chapters of this book, we have gathered a variety of examples demonstrating how such data stories can be created and taught to new generations of data scientists and, most especially, those who aspire to using the wealth of data available yet who lack as strong a background in data science and analytics. It is this group—those we call data translators—who navigate the shifting boundaries between disciplinary expertise and data science, interpreting data in meaningful and accurate ways for others.

How can we teach new generations of data translators to navigate these boundaries? The case studies in this book suggest three factors need attention. The first is that to develop effective data translators we must acknowledge interdisciplinarity as

B. Samuels (✉)
Faculty of Kinesiology and Physical Education, University of Toronto, Toronto, Ontario, Canada
e-mail: boba.samuels@utoronto.ca

D. Kotsopoulos
Faculty of Education, University of Western Ontario, London, Ontario, Canada
e-mail: dkotsopo@uwo.ca

D. G. Woolford
Department of Statistical and Actuarial Sciences, University of Western Ontario, London, Ontario, Canada
e-mail: dwoolfor@uwo.ca

a guiding force in data science. In other words, data science is not just for data scientists. As this book demonstrates, scholars and stakeholders across many different fields are teaching and learning how to tell data science stories for their audiences. However, this ubiquity in the use of data science and analytics methods also runs the risk of incorrect analyses leading to incorrect conclusions—just because someone has data that can be run through a data modelling method, doesn't mean that it is being done correctly. This was highlighted by Li in Chap. 10 through an analysis of binary (correct/incorrect) response data that demonstrated the need to understand the underlying assumptions of data modelling methods and to select an appropriate technique accordingly. Hence, more attention to the ways in which such teaching occurs across disciplines, and particularly when such attempts are multidisciplinary, is needed. Drawing on the expertise and experiences of different stakeholders in this endeavour is good practice. The expertise required to generate big data, for instance, does not lead directly to the expertise needed to use this data. Even with so much data available, only 18% of companies can use the data they generate effectively [1]. More is needed, therefore, than technical skills in data science to effectively use data to address real discipline-specific or workplace issues.

In Schanzer et al.'s Chap. 3 description of public health policy work, multidisciplinary committees are used to shape the creation of policy guidelines, demonstrating that approaches that include multiple different perspectives can be effective. They suggest that teaching using class discussion to identify multiple interpretations of guidelines can raise students' familiarity with such guidelines as well as the critical review process. Such knowledge and skills are necessary in understanding how data analysis is used to support creation of the guidelines. Creating opportunities for interaction between disciplinary experts and creating classroom models that replicate the sharing of different perspectives are promising approaches for encouraging the critical review of data analysis necessary for ensuring high quality data interpretations.

The second factor necessary for teaching people to become effective data translators is to identify a framework for knowledge exchange (KE) between stakeholders. In discussion of KE in the wildland fire management context, McFayden et al.'s Chap. 12 describes the need for intentional engagement in the sharing of knowledge and proposes that a KE framework can facilitate such iterative engagement between researchers and practitioners. In their words, "Effective KE in fire management helps ensure that real-world problems are understood by researchers, the research is relevant, and the results are integrated into fire management practices." They describe how an active learning approach in the classroom can facilitate use of such a KE framework by students who may go on to become data translators. In this approach, which is facilitated by knowledge brokers, students practice the behaviours and skills relevant to address a typical real-world problem by actively engaging with multidisciplinary researchers and practitioners rather than focusing on the technical requirements of data analysis.

In their example, McFayden et al. describe a scenario in which students work in teams with an external "client" with whom they meet to address the client's problem. The class instructor acts to facilitate communication between the students and client by brokering knowledge between the stakeholders, working with the client to guide students through the KE framework. Students take on different roles in their team depending on their background and technical skill level. Various modes of communication are practiced including email exchanges, phone/video meetings, progress reports, presentations, and a final report; peer feedback is also used to encourage the sharing of multiple perspectives and develop the ability to provide constructive criticism. Such an active learning approach provides students with an opportunity to not only practice essential skills necessary for data analysis, but also to experience how a KE framework can guide them to integrate such skills with behaviours conducive to developing mutually beneficial relationships with clients and peers. Identifying and applying such a KE framework is thus likely to be productive for data translators.

Teimourzadeh and Kakavand in Chap. 7 present a staged process for students to learn how to define and address a business problem using data. The "framework" for this staged process is the data analytics workflow with students using two Tableau apps to develop data literacy and data translation skills. The authors describe an "experiential" learning approach in which students practice—as in the active learning approach described by McFayden et al. –the skills necessary for real world use. Instructors are provided with a step-by-step plan that focuses on helping students use freely available software apps to move through six stages of a data analytics workflow, from identification of a business problem, through data collection, processing, analysis, and visualization, to data translation. This approach may be particularly helpful for students with little background in data analysis because it presents them with a framework for their activities that they can apply across a range of possible contexts as well as experience in applying this framework.

Finally, the third factor identified in these chapters' case studies is language calibration, namely the attention to language in flexible ways to carry out effective data translation between participants. Zwiers and Zhang's Chap. 1 discusses the calibration of language related to climate change by an intergovernmental organization, and shows how such development of consistency in use of calibrated terms enables confidence in users and an ability to demonstrate progress over time. They describe, for instance, how specifying the probability related to various terms (e.g., *virtually certain*, *likely*, *very unlikely*) has enabled stronger claims to be stated about scientific conclusions. Most significantly, such attention to clear communication through development of consistent, calibrated terminology provides policy makers and scientists a tool to share information on climate change with confidence that uncertainties in meaning are minimized. The pedagogical implications of this work include drawing students' attention to the value that data translators must place on precise language and the internal values of a discipline or organization. In this case, the need for climate scientists and organizations to avoid confusion and emphasize consensus is necessary to avoid attacks on their credibility in the contentious context of climate change debates.

Falconer takes a wider view on language in Chap. 11, focusing on identifying the rhetorical strategies used when data created in one context is taken up and transformed for use in another domain. By distinguishing between recontextualizing knowledge and data translation, with the former describing a process of transforming knowledge or data from one form to another, while the latter refers to making knowledge accessible (i.e., translated) for different audiences for different purposes, Falconer describes how multidisciplinary and multisectoral panels transform scientific data for use by governmental departments for the purposes of policy development. Falconer identifies strategies including the visual representation of people, numerical representations, and written explanations as some of the means by which data is transformed for later use. In the classroom, these strategies can be identified, analysed, and taught to students who envision themselves in roles doing such "boundary work."

These three factors—interdisciplinarity, a knowledge exchange framework, and language calibration—necessary for effectively teaching people to become data translators reflect issues familiar to writing instructors working to teach students how to write in the disciplines. A well explored concern, for instance, is how to enculturate novice students to a discipline so they can increasingly participate and be seen as legitimate actors within that discipline. For example, how can we teach students to think and write like a scientist? For students new to a discipline, the conventions, genres, language, and values of that discipline are largely unknown and invisible. Making such disciplinary elements visible and enabling practice to use these elements incrementally and appropriately allows students to begin identifying as disciplinary insiders, a process known as legitimate peripheral participation [8]. This process is relevant not only to novice students, but to those who are established within disciplines and who now must interact with and adjust to other disciplines to take advantage of the expanded opportunities available through interdisciplinarity.

This discussion of conventions, genres, values, and participation in a disciplinary community is central to the issue of how to become a data translator. Being a data translator requires that individuals participate and are perceived as legitimate actors within that discipline, not as outsiders. Providing students with experience conducting data analyses using authentic data and then translating that data for use within a specific discipline to address a discipline-specific question teaches students about the conventions that must be followed, the genres that are possible, and the values that require acknowledgment. Students thus learn what it means to be a disciplinary insider and are therefore able to communicate appropriately with disciplinary experts. It is this ability to communicate in discipline-appropriate ways that makes them translators.

Translation is not simply about participants speaking different "languages"—one speaking with technical terms from data science and analytics and one speaking with discipline-specific terminology. Rather, when complex problems intersect with large and complex data structures, it is likely that neither the background knowledge, skills, epistemological values, nor vocabulary of the various stakeholders are aligned, with the result that communication even in a common language would be challenging. Data translators therefore bring together an immense variety of skills crucial to the success of an organization using data, including their data analytics skills, technical fluency, communication skills, project management, and an intimate knowledge of how the data can address organizational barriers to advance the companies' or organizations' interests [6].

Just as expert translators of poetry or novels from one language to another are highly valued for their insights into both languages and both cultures, so too are data translators increasingly valued for their ability to bridge differences between data science and other disciplines or institutions. The specific contributions of data translators have been noted, for instance, in the workplace [7]. For example, Marr (2018), writing for Forbes, went so far as to say that due to their ability to distill complex information to make it understandable to decision makers, employers should hire data translators rather than data scientists. Data translators, in other words, are those who can bridge the technical-stakeholder gap [9]. The hiring of those who can tell stories with data may thus make up the next big wave of hiring [5].

While it is too soon to tell if such predictions prove accurate in the workplace, in the academy there are increasing signs that the need to disseminate information to non-experts is well established and growing. Students and scholars are, for instance, increasingly required to write plain language abstracts for research grant applications and publications, with such translation work valued as an integral component of the research process. Though such tasks may not involve complex data structures or data science, they demonstrate the need to learn how to tell accurate and engaging stories to those without expert knowledge, making knowledge interpretable. In the workplace, data translators help to bridge the "interpretation gap", providing the insights from large data sets to make this information actionable to less-technical workers [4]. While the utility of data translators has been noted in STEM fields such as sustainable development [2] and biomedicine, much of the focus on data translation in academic disciplines is on translating data in underlying models rather than on the training of individuals to be data translators [3].

In this collection, we have shared a number of data science stories to describe early approaches to the pedagogy of data translation. Many other stories exist across a multitude of disciplines and fields, stories of instruction undertaken perhaps tentatively at first, but then with increasing confidence. These stories may provide models for future instruction. By bringing together experts who can demonstrate how such instruction is possible, we hope to encourage other instructors to share their approaches, with the goal of creating a robust and flexible pedagogy for future data translators.

References

1. Ariker, M., Breuer, P., McGuire, T.: How to get the most from big data. McKinsey Inst. (2014). https://www.mckinsey.com/capabilities/mckinsey-digital/our-insights/how-to-get-the-most-from-big-data
2. Auad, G., Fath, B.D.: Towards a flourishing blue economy: Identifying obstacles and pathways for its sustainable development. Curr. Res. Environ. Sustain. **4**, 100193 (2022)
3. Biomedical Data Translator Consortium: The Biomedical Data Translator program: conception, culture, and community. Clin. Transl. Sci. **12**(2), 91 (2019)

4. Brady, C. Forde, M., Chadwick, S.: Why your company needs data translators. MIT Sloan Manag. Review. (2016). https://sloanreview.mit.edu/article/why-your-company-needs-data-tra nslators/
5. Fisher, A.: Now hiring: People who can translate data into stories and actions. Fortune. (2019). https://fortune.com/2019/10/12/human-intelligence-data-translation/
6. Henke, N., Levine, J., McInerney, P.: You don't have to be a data scientist to fill this must-have analytics role. Harv. Bus. Review. (2018). https://hbr.org/2018/02/you-dont-have-to-be-a-data-scientist-to-fill-this-must-have-analytics-role
7. Kenney, A.: 6 steps to become a finance data 'translator'. Financ. Management. (2021). https://www.fm-magazine.com/news/2021/oct/6-steps-become-finance-data-translator.html
8. Lave, J., Wenger, E.: Situated learning: legitimate peripheral participation. Camb. Univ. Press., Cambridge, UK (1991)
9. Maynard-Atem, L., Ludford, B.: The rise of the data translator. Impact **1**, 12–14 (2020)

Printed in the United States
by Baker & Taylor Publisher Services